WHAT IS SCIENCE?

BY

NORMAN CAMPBELL

SC.D., F.INST.P.

DOVER PUBLICATIONS, INC.

NEW YORK

This Dover edition, first published in 1953, is an
unabridged and unaltered republication of the work
originally published by Methuen & Co., Ltd. in 1921.
This edition is published by special arrangement
with Methuen & Co., Ltd.

International Standard Book Number: 0-486-60043-2
Library of Congress Catalog Card Number: 53-6948

Manufactured in the United States of America
Dover Publications, Inc.
180 Varick Street
New York, N. Y. 10014

CONTENTS

WHAT IS SCIENCE?

CHAPTER I

THE TWO ASPECTS OF SCIENCE

THERE are two forms or aspects of science. First, science is a body of useful and practical knowledge and a method of obtaining it. It is science of this form which played so large a part in the destruction of war and, it is claimed, should play an equally large part in the beneficent restoration of peace. It can work for good or for evil. If practical science made possible gas warfare, it was also the means of countering its horrors. If it was largely responsible for the evils of the industrial revolution, it has already cured many of them by decreasing the expenditure of labour and time that are necessary for the satisfaction of our material needs. In its second form or aspect, science has nothing to do with practical life and cannot affect it, except in the most indirect manner, either for good or for ill. Science of this form is a pure intellectual study. It is akin to painting, sculpture, or literature rather than to the technical arts. Its aim is to satisfy the needs of the mind and not those of the body ; it appeals to nothing but the disinterested curiosity of mankind.

The two forms, practical and pure science, are probably familiar to everyone ; for the necessity for both of them is often pressed on the public attention. There is sometimes opposition between their devotees. Students of pure science denounce those who insist on its practical value as base-minded materialists, blind to all the higher issues of life ; in their turn they are denounced as academic and unpractical dreamers, ignorant of all the real needs of the world. If the two forms of science were really inconsistent with each other, both sides could present a strong case. Few would deny that, in some sense, intellectual interests are higher and more noble than material interests ; for it is in the possession of intellectual interests that we differ from the brutes. Indeed, it may be urged that the only reason why men should care for their material interests, or why they should care to obtain anything but mere freedom from the pains of cold and starvation, is that they may have the leisure and the freedom from care necessary to cultivate their minds. All but the most base must have respect, if not sympathy, for those who prefer to live laborious days in the pursuit of pure learning rather than to devote their energies to the attainment of personal wealth and ease. But to press this point of view is to misrepresent the issue. More than the interests of the student himself are involved ; and though the benefits of pure and abstract science may be higher than those of practical and useful science, they are much less widely distributed. It is only a small minority of mankind who can hope to share the former ; few have the mental equipment necessary for the full enjoyment of the quest and discovery of pure knowledge ; and of these few not all are able to undertake the long and strenuous training that is a necessary preliminary to full enjoyment. On the other hand the benefits of practical science might be shared—even if they are not shared in our present society—by almost every one ; the vast majority do not possess the freedom from material cares necessary for the full development of their higher interests ; and if practical science can so facilitate the satisfaction of material needs as greatly to increase the number who have that freedom, its value, even if judged by the least material and the most academic standards, may be in no way inferior to that of the purest and most abstract learning.

However, to-day it is probably unnecessary to pursue such arguments. For it is now generally recognized that the two forms of science, whatever may be their relative value, are in fact inseparable. The practical man is coming to understand that the earnest pursuit of pure science is necessary to the development of its practical utility, though he may sometimes have strange notions of how that pursuit may best be encouraged. And academic students are finding that the problems of practical science often offer the best incentive to the study of pure science, and that knowledge need not be intellectually uninteresting because it is commercially useful. In a later chapter we shall consider in rather greater detail what is the relation between pure and practical science and why they are so inseparable ; but it is well to insist at the outset upon their close connexion. For the distinction between the two has undoubtedly discouraged the study of science among the W.E.A. classes for which this little book is intended primarily. Those who are more familiar with the practical aspect are apt to think that the study of science can be nothing but a disguise for technical and vocational education ; while others think that anything so entirely abstract as pure science can have no bearing on the practical problems of society in which they are more directly interested. Both views are entirely mistaken ; the study of science need be no more " technical " than the study of music, and, on the other hand, it may be quite as practical as that of political economy.

Nevertheless, though pure and practical science are inseparable and merely different aspects of the same study, it is necessary to remember the difference between them. And I want to point out here, once and for all, that what we are going to study directly is pure science ;

that the motive of our study is supposed to be intellectual curiosity without any ulterior end ; and that our criterion will be always the satisfaction of our intellectual needs and not the interests of practical life. This procedure would be necessary even if our ultimate concern were rather with practical science. For it is only if we understand the nature of pure science that we can interpret with confidence the knowledge that it offers and apply it rightly to practical problems. Science, like everything else, has its limitations ; there are problems, even practical problems, on which science can offer no advice whatever. One of the greatest hindrances to the proper application of science to the needs of the community lies in a failure to realize those limitations ; if science is sometimes ignored, it is often because it has been discredited by an attempt to extend it to regions far beyond its legitimate province.

But it may be said, if the appreciation of pure science must always be confined to a few serious students, what is the use of such an attempt as this to make it intelligible to the plain man ? The answer is simple : I only said that the *full* appreciation must be so confined. Nobody can appreciate good music to the full unless he has trained himself by careful study, yet most of us can get some value from a concert ; perhaps we get more actual enjoyment than a skilled musician. It is just the same with science. Indeed, there is little doubt that science is the easiest of the branches of pure learning for the amateur. It is quite common to find men of high intellectual gifts and not without learning, who are perfectly incapable of understanding what mathematics or philosophy is all about, why anybody should ask such absurd questions and how they think anyone is the better for the answers they give. A similar complete indifference to science is not common ; almost every one can be made to understand what science is about, and almost every one derives some satisfaction from the answers which it offers. This wider appeal is often attributed to the practical interest of science, but that explanation cannot be the whole truth ; for some scientific doctrines, such as the Copernican theory and the theory of evolution, have convulsed society without having the smallest effect on anybody's material comfort. The true reason is easy enough to discover, but its complete discovery would answer most of the questions which we are going to ask.

THE DEVELOPMENT OF PURE LEARNING

The main question which this book is designed to answer may be expressed simply : What is Science ? We have already answered it partly in saying that science is a branch of pure learning which aims at intellectual satisfaction. But it is not the only branch, and we must ask next what it is that distinguishes science from other branches. Is the distinction in the subject-matter that it studies, or in the manner in which it studies it, or both together, or, possibly, something quite different ? The formal answer that I propose to give could be given at once and quite briefly, but since at first sight it might not appear plausible or even intelligible, we shall do better to lead up to it more gradually.

All branches of pure learning spring from a common stock. We generally think of " pure learning " as something peculiarly characteristic of the highest state of civilization and as something which could develop only when man had advanced a very long way from savagery. But as a matter of fact the instinct which inspires pure learning is one of the oldest and the most primitive ; man begins to seek answers to the riddles which still perplex the most abstruse of philosophers before he begins to wear clothes or to use metal implements.

Whether we regard the childhood of the race or of the individual, we find that, as soon as man begins to think at all, he utters his perpetual question, Why ? The world around him does not appear to him immediately intelligible ; it seems to have no meaning and to be arranged on no comprehensible plan. He asks how the world came to be what it is and why it is what it is. To such questions, inspired in the first place by mere curiosity rather than by a desire to control the world to his liking, answers of some sort are given by the most elementary religions and the crudest systems of magic. Some form of religion or magic, which attempts to explain the world in terms of ideas that are the product of thought and reflection rather than of immediate perception, seems characteristic of almost all races of men, however low their intelligence and their material advancement.

It is, of course, impossible to determine certainly whether these rudimentary attempts at pure knowledge, which are found among the less developed races of to-day, represent different stages in an evolution through which all men's ideas have passed and must pass, or whether they are entirely independent. And in particular it is impossible to trace back the history of our own pure knowledge to its earliest origins. But we can trace it back a very long way to the speculations of the ancient Greeks in the third and fourth centuries before our era. Greek thought, in the earliest stage in which we encounter it, is very different from the primitive religions and magics of savages ; but classical scholars find in it relics which lead them to believe that its first origins were not very different from the ideas of the most backward races of the present day. But in spite of these relics, the advance that was made in the great Age of Greece was enormous. It has largely determined all subsequent European thought ; and it is not too much to say that there was less advance made in pure learning in the 2,000 years from 300 B.C. to A.D. 1700, than in the 200 years from 500 B.C. to 300 B.C. All speculation on the nature and meaning of the world throughout the Roman Age of civilization, through the Dark Ages and through the Mediæval Age, drew its inspiration directly from the Greek philosophers, and especially from Aristotle ; it is not until the Renaissance is well advanced that a new stream enters from a wholly independent source. And even to-day, when there is no school of thought which maintains the Greek tradition in anything approaching purity, its influence is still potent. Its effect upon language is still most evident ; we cannot speak upon any abstract subject, or express any general idea, without using words which are either Greek or direct Latin translations of Greek words. And since words are an indispensable instrument of thought, in using Greek words we are bound to be influenced to some extent by Greek ideas.

Now Greek learning formed a single whole. To-day we distinguish many branches of learning—mathematics, science, philosophy, history, and so on. But this division is quite modern ; Greek thought made hardly any distinction between them. (Perhaps an exception should be made of history, and also of the study of languages : the Greeks did not study languages ; they knew none but their own.) Even at the beginning of the nineteenth century, all learning was called philosophy or (less frequently) science, and a man was called a philosopher even if he studied what we should now call mathematics or science. Until well on in that century the universities recognized only one form of study as a means to a degree, and that form included a little of most of the forms recognized, and sharply divided, at the present day. The reason is not to be found simply in the smaller body of knowledge at that time, so that one mind could grasp all that was known ; there was a real absence of distinction between what are now regarded as different kinds of knowledge. Our ancestors would have strenuously denied that a great mathematician could be ignorant of philosophy or a great philosopher ignorant of science. One of the widest differences between modern and ancient thought is the recognition that there are independent systems of thought and independent bodies of knowledge, and that errors in one branch are not necessarily accompanied by errors in another.

SCIENCE AND OTHER STUDIES

Of course the branches into which pure learning has separated have been changed greatly since, and in virtue of, their separation. None has been more affected in this manner than science ; the great development of science of the last century is intimately connected with its divorce from philosophy. And the changes are so great that it is perhaps hardly right to regard the science of to-day as the same thing as the science which was not distinguished from other studies in Greek and mediæval thought. Nevertheless this discussion has not been irrelevant ; for it reminds us that science, like all other attempts to satisfy the curiosity of man, has its ultimate roots in the simplest and most instinctive speculations. It shows us also that, however distinct from all other kinds of pure learning the science of to-day may appear, the exact line of division and the exact criterion are likely to be difficult to lay down ; a distinction that was overlooked for 2,000 years is not likely to be discoverable by a casual inspection. Again it suggests that, since the separation of science has taken place in times so recent, one way to discover the distinction may be to inquire into its history of the word.

This history is quite simple. When it was recognized that the studies which now form part of science required a separate name, they were called " natural philosophy " in distinction to " moral philosophy " ; and they were also called " natural science " in distinction to " moral science " ; for at that time " philosophy " and " science " had practically the same meaning and were used interchangeably, although the former was the commoner. All these expressions survive ; at the older universities a professor of natural philosophy is indistinguishable from a professor of physics or chemistry ; and " moral science " is a common name for what is more usually called philosophy. That " natural philosophy " has become almost obsolete while " natural science " survives, is due partly to the inexplicable vagaries of language which determine, apparently at random, which of two synonyms is to die out ; but it is also partly due to the fact that the older branches of learning from which the students of science desired to separate themselves were more often known as philosophy than as science. Again the " natural " has been dropped, and only the " science " retained, partly by mere abbreviation (just as " omnibus " has been changed into " bus "), and partly because students of science were by no means averse from hearing their study called " science " without any qualification ; for " science " is simply the Latin for " knowledge," and the implication that all that is not science is not knowledge, naturally flattered their vanity. And it is important to remember this history. For the older and more general use of the word to mean pure knowledge in general, or indeed any kind of knowledge, has not vanished ; and we must be on our guard against imagining that everything to which the words " science " and " scientific " are attached to-day have anything more to do with natural science than with any other kind of knowledge. When a journalist speaks of a " scientific batsman " he merely means that he is skilful and does not imply that he is learned in physics or astronomy.

Here no doubt the more general use is clearly distinguished from the more special, but some misunderstandings about the science that we are going to consider probably arise from this double use of the word.

SCIENCE AND NATURE

But why were these special branches of learning called " natural " ? Not because they were more natural, in the conversational sense, than any other, or even in the Shakespearean sense (which means idiotic) ; but because they were regarded as being especially concerned with nature. And what is meant by " nature," and how is science especially concerned with it ? The term " nature " has never been used in a very precise sense capable of accurate definition, but it seems generally to be employed in contradistinction to man ; nature, we may say roughly, is everything in the world that is not human. Nature is regarded as the antagonist of man, the obstacle which he has to overcome and the enemy he has to fight, although he may sometimes turn the enemy into a friend by judicious action. This idea will be found, I think, to underlie most uses of the word. It is true that sometimes, and more particularly in the middle of the last century, man has been regarded as part of nature ; for instance, one of Huxley's best-known books is called " Man's Place in Nature " ; but the view that man was part of nature was felt to be rather hetero-dox and startling, an overthrowing of many preconceived beliefs ; indeed, the phrase was used by Huxley largely in order to challenge accepted opinion.

Again, the opposition of nature and man is reflected in the terms used to distinguish the branches of pure learning which were most clearly separated from science. They were termed " moral " philosophy or science. Now " morals," even in the very general sense attributed to the term when used in this connexion, are particularly human. Common sense divides the world into three great divisions—man, animals and plants (or living beings other than man), and inanimate objects. To the third division the idea of morals is clearly inapplicable, whether it refers to all mental processes or more particularly to right conduct ; and it is applicable only in a very limited degree to the second ; the first is its proper province. The distinction between natural and moral philosophy suggests at once that the latter is concerned especially with man and his ways ; the former with everything that is foreign and external to man. Nature means practically the part of the world which man regards as external to himself.

Accordingly it is suggested that science should be defined as that branch of pure learning which is concerned with the properties of the external world of nature. Its business is to find out accurately what those properties are, to interpret them, and to make them intelligible to man ; the intellectual satisfaction at which it aims would be secured completely if this external world could be reduced to order and be shown to be directed by principles which are in harmony with our intellectual and moral desires. On the other hand, science will not, on this view, be concerned with anything distinctively human ; it will not consider human thoughts and actions, ask what those thoughts and actions are, or examine and criticize them. And this suggested definition of science would probably have been accepted very generally at the time when science was first distinguished from other branches of learning under the name of natural philosophy. Nevertheless, there are difficulties in accepting it. For, according to the view that has been put forward, all pure learning arose ultimately from man's desire to understand the world ; it was his opposition to the external world of nature that started his inquiry and his search for explanation. If, then, it is this external world which is the special province of science, we should expect to find that learning would become more distinctively scientific (in the modern sense) as we trace it back through the ages, and that branches, other than science, which are now separated from the common stem, would appear at only a relatively late stage in the growth. Actually, of course, we find exactly the opposite ; what is now recognized as science, as the study of nature and the external world, is the youngest and not the oldest of the departments of pure learning. Again, there are undoubtedly studies, usually accepted as sciences, which specifically deal with man and not with the external world which is contrasted with him ; psychology and anthropology are examples ; how are they consistent with the view that science is characteristically non-human ? Lastly, it is generally recognized to-day that science differs from other branches, not only in the subject-matter that it studies, but also in the manner in which it deals with this subject-matter. Even if we could define the subject-matter of science as being the external world of nature, we should still be left with the inquiry, which is really more interesting, why the difference in the subject-matter involves so great a difference in the attitude towards it.

SCIENCE OR SCIENCES ?

These difficulties show that we cannot obtain the answer that we require to our question, What is Science ? by simply accepting the answer that might have been given a hundred years ago. On the other hand, it is indubitable that this answer is part of the truth. To that inquiry we shall proceed in the next chapter, and with it shall start the serious part of our discussion. But before we proceed, we shall do well to consider very briefly one other matter which belongs properly to this preliminary stage. Are we right to speak at all of " science " ? Every one knows to-day that there is not one science but many. Physics, chemistry, astronomy, geology, zoology, botany, physiology, psychology, and so on, although all called " sciences," seem to be branches of knowledge almost as separate as any science is from philosophy. A chemist may be as ignorant of botany as a philosopher of mathematics. Can we say anything that is true of all these sciences and is not equally true of mathematics or philosophy ? Well, that is one of the questions that we have to answer, and our answer will be affirmative ; we shall lay down a criterion which appears to distinguish all sciences from any other branch of pure learning. But a word may be said here about the relations of the different sciences.

The division between them corresponds in part to the crude common-sense division of the external world of nature. Thus we find some sciences (zoology, botany, physiology) dealing with living beings and others (physics and chemistry) with inanimate " matter." Further we can distinguish sciences which deal with particular objects from those which deal with the common substratum of objects. Thus geology deals with one particular object, the earth ; and astronomy with other particular objects, the stars ; zoology and botany consider particular animals and plants. On the other hand physics and chemistry deal with the substances of which all particular material objects are composed ; physiology with the functions common to all living beings. So far the divisions between the sciences lie along the lines that we should expect if science is the study of the world of nature. But such divisions can only be made very roughly. The province that is actually regarded to-day as belonging to each science is very largely the result of historical accident ; one line of inquiry leads to another, and a new line of inquiry is often assigned to the science that was the particular study of the first investigator of that line, without discussion whether the allocation can be justified on any formal principle.

Such considerations clearly justify the view that science is a single whole and that the divisions between its branches are largely conventional and devoid of ulterior significance. But, though science may be really one, its range and complexity to-day is so great that the most learned of mankind cannot profess to a serious knowledge of any but a very small part of it. And therefore perhaps I ought to justify and explain my temerity in writing of science in general. I should point out that physics is the only science of which I profess an expert's knowledge, and that the discussion is bound to be directed from the standpoint of a student of that science. But it is generally admitted that physics is in some sense more fundamental than any other science, and that the results of physics constitute, in some sense, the starting point of other sciences. Why there should be that relation is a matter for subsequent inquiry; but the admitted fact of the relation makes it certain that, if we decide what is physics, what is its fundamental subject-matter and its method of dealing with it, we shall have gone a long way towards answering similar questions which may be raised concerning any other branch of science.

However, there is one question which should be noted here. The examples of the various sciences that have been given include none of the studies that lie on the border line. Every one is prepared to grant that botany and chemistry and physics are properly called sciences, though there may be some doubt exactly what they have in common ; but there are two studies of wide interest the claims of which to be sciences are not universally admitted. I refer to history and economics. The judgment on these claims cannot, of course, be properly passed until our inquiry into the characteristics of science are further advanced ; but it will be convenient to anticipate some of our conclusions in order to dismiss the matter. When he has read the two following chapters, the reader should consider the question for himself.

The view to which I incline is that history cannot be usefully grouped with the characteristic sciences, and the reason will appear at once in Chapter III. The main concern of history is not with laws, but with particular events. The decision concerning economics is more difficult. A civilized community is part of " nature " and there is no reason for thinking that such a community may not be subject to laws in the scientific sense. But I have very grave doubt whether any economic " laws" hitherto enunciated are laws in that sense ; and the basis of my doubt will appear in the next chapter. Economics might be, and some day may be, a science ; but at present it is not. That is my opinion ; but as I profess no special knowledge of economics, it may easily be wrong. But I think it is certain that economics, whether or no it is a science, is so different from those that we are going to consider that it would be rash to apply to it any of the conclusions that we shall reach.

CHAPTER II

SCIENCE AND NATURE

WHY DO WE BELIEVE IN AN EXTERNAL WORLD ?

HOW do we come to have any knowledge at all of the external world of nature ? The answer is obvious. We learn about the external world through our senses, the senses of sight, hearing, and touch, and, to a less degree, those of taste and smell. Everything that we know about the external world comes to us from this source ; if we could neither see, hear, nor feel, we should know nothing of what was going on round about us, we should not even know that there was anything going on round about us ; we probably should not even form the idea that there is such a thing as the external world.

So much is clear and indubitable. But now we have to ask a much more difficult question, and one concerning which there has been much more difference of opinion. Why do we regard our senses as giving us knowledge of the external world ? Every one agrees that *if* we have any knowledge of such a world, it is derived from what we see, hear, and feel, and not from any other source ; but it is quite possible to doubt that what we see, hear, and feel, does really give us that knowledge, or that we are right in interpreting the evidence which we derive from our senses in the manner in which we do habitually interpret it. It is rather difficult for those who are unfamiliar with the controversies that have raged round this matter to grasp the position of those who express such doubts ; it seems to us so obvious that when we hear a noise or see an object we are perceiving something external to ourselves. And the difficulty of grasping the position is intensified because all our habitual language is based on the assumption that there is an external world which we perceive. For instance, when I want to call attention to the sensation of sound, I can only say that "I hear a noise"; but the very form of the words which have to be used to convey my meaning imply that the "noise" is something different from the "I," who hears it. Nevertheless it is necessary to try to understand how such doubts can be put forward.

They are based on the fact that the experience of seeing an object or hearing a sound is an event which takes place in my mind; it is a kind of thought—if we use the word "thought" to mean anything that goes on in my mind. That fact is expressed when it is said that ' I " hear the noise. Though the noise may be the same, the fact that "I" hear it is different from the fact that "you" hear it; the first fact is something that happens in "my" mind, the second something that happens in "your" mind. The noise, or the thing that causes the noise, may be something in the world of nature, external to both you and me; but the hearing of the noise, which is the fact on which you and I base the conclusion that there is a noise or that there is an external object making a noise, that hearing is not something external; it is something internal to you or to me, according as you or I hear it. This view, that the perception of an external object is something internal to the person who perceives it, is as much part of the common-sense attitude towards the matter as the view that the perception gives evidence of an external object.

But now we may argue thus. It is agreed that the perception of an external object is something internal to the perceiver; it is one of the thoughts, or the mental events, of the perceiver. On the other hand we do not regard *all* thoughts of a perceiver as giving evidence of an external world; there are also thoughts which are purely internal and totally unrelated to the external world. Indeed, it is such thoughts which give rise to the idea of a perceiver who perceives the external world. For I regard all my perceptions as "my" perceptions, because they are all connected together by thoughts of other kinds. Thus I can remember my perceptions and call them to mind ; I can think about them and compare one with another ; I can judge that they are pleasant or unpleasant and desire that some and not others should recur. These thoughts about my perceptions I regard as characteristically internal to me ; they are just the things which make up " me " ; once more, it is these thoughts about my perceptions which make me regard them as " my " perceptions. It is very difficult to convey these sentiments in words, just because, as has been said already, all words assume these sentiments. But I hope that any reader who considers the matter will agree that the conception of an external world, which I perceive, is founded as much on the idea that there are thoughts which are wholly part of me, and have nothing to do with the external world, as on the idea that there are other thoughts which, though they are also part of me, are intimately connected with the external world, and inform me of that world.

If this view can be grasped, the basis of the doubts that we are considering becomes clear. Some of my thoughts I regard as wholly internal ; others, forming the special class of sensations or perceptions received through the organs of sense, I regard as rather part of the external world. Why, it may be asked relevantly, do I make this distinction ? If it is necessary to regard some of my thoughts as wholly internal and giving evidence about me and not about the external world, why do I not so regard all my thoughts ? If one class of my thoughts does not give any information about an external world or even any evidence that there is such a world, why do I regard another class as giving such information and evidence ? Is it not at least reasonable to regard all thoughts in the same way, and to dispense altogether with the recognition of an external world as the cause of some of my thoughts ?

No serious school of thought has seriously maintained the position indicated in these questions. Indeed, to maintain it or to argue about it would be impossible, or extremely foolish, for anyone who believed it. For— we shall revert to this point in a moment—if there is no reason for believing in an external world, there is no reason in believing that there are other people with whom to argue or against whom to maintain a position. The view that has been based on the contention that sensations are only thoughts, and therefore, like all other thoughts, internal rather than external, is not that sensations give no evidence at all for believing that there is an external world, but only that the information which we derive from our senses about the external world is not so simple and direct as we often imagine, and consequently, that our first impressions about the external world may be very far from the truth. However, for our purpose it is necessary to press the more extreme view, and to ask why we distinguish so sharply between sensations and other thoughts, and why we regard the former and not the latter as giving evidence of, and information about, an external world. In pressing the view, I have, of course, no intention of maintaining that our habitual distinction is not valid ; I only want to elicit what is the difference between the two classes of thoughts which makes it valid. Our question is, What is the difference between the thoughts which we call sensations, and connect with our organs of sense, and the thoughts which we call memory, or reasoning, or will ; and why does this difference lead us to refer the first class, but not the second, to an external world ?

THE CHARACTERISTICS OF SENSE-PERCEPTIONS

There are two such differences. In the first place, our sensations are much less under our control than are our other thoughts ; in the second place, other people agree with us in our sensations far more than they agree with us in our other thoughts. That is, in brief, the answer which I propose to give to the question ; it must now be explained and expanded.

The first distinction is that our sensations are less under our control than our thoughts. They are not wholly beyond our control ; for, if I close my eyes, I can refuse to see, and if I do not put out my hand, I can often refuse to feel. But if I do look at an object it is wholly beyond my control whether I see that it is red or see that it is blue ; and if I put my hand into the fire I cannot help feeling that it is hot and not cold. On the other hand, thoughts, other than sensations, are not wholly under control ; I cannot always remember what I want to, and I cannot always keep my attention on my work ; even my will is sometimes not under control, and I may feel that there is a conflict within me. But, though in this matter our sensations and our other thoughts may differ in degree rather than in kind, it will probably be recognized that there is this difference, and that it is part of the reason why we feel that our sensations are intimately connected with something external and do not take their origin wholly within ourselves. For what is not under my control is not really part of me ; what I mean by " me " or " myself " is simply what is under the control of my will : my will is myself. (Of course this is one of the statements which it is impossible to express accurately in language which assumes the position which is under discussion.) I recognize this fact most clearly in those curious cases when there is an internal conflict of will, and when " my " will seems divided against itself; I then speak of the antagonistic wills as if they were those of two different persons. If I act in a way which is contrary to my normal will, I say that " I was not myself." This feeling that " I " am practically indistinguishable from my will, and that what is not subject to my will is not me, is undoubtedly one of the main reasons for referring sensations, which are often wholly independent of my will, to a foreign and external world. It has also, as we shall see, a bearing on the second and more important difference between sensations and other thoughts, to which we must turn next.

This second difference is that other people agree with me much more closely about sensations than they do about any other kind of thought. The fact, expressed in that manner, is extremely familiar. If I am in a room when the electric light bulb bursts, not only I, but everyone else in the room (unless some of them are blind or deaf), hears the explosion and experiences the change from light to darkness. On the other hand, apart from sensations, we may all have been thinking about different things, remembering different things, following different trains of reasoning, and experiencing different desires. This community of sensations, contrasted with the particularity of other kinds of thoughts, leads naturally to the view that the sensations are determined by something that is not me or you or anybody else in the room, but is something external to us all; while the other thoughts, which we do not share, are parts of the particular person, experiencing them. This simple experience is probably the main reason why we have come to believe so firmly that there is an external world and that our perceptions received by our senses give us information about it.

For this is the test which we apply in practice when any doubt arises if we are experiencing sensations which give

information about the external world. In general we
have not the slightest difficulty in distinguishing such
sensations from other thoughts and other mental events ;
but there are exceptions to the general rule. Thus when
we awake from a very vivid dream there is often a considerable interval in which we are not sure whether our
dream experiences were real ; we have been experiencing
perceptions so very like the sensations which tell us about
the external world that, if we were left to judge solely
on the basis of their mental quality, we should probably
think they were sensations and gave us information
about the external world. Doubtless we have all had so
many dreams and are so well aware of the circumstances
in which they are likely to occur that a very brief reflection is usually sufficient to enable us to decide whether
we were dreaming or were really hearing or seeing something. Nevertheless doubtful cases do arise ; and, when
they do arise, what do we regard as a certain test to
decide the matter ? Surely the test is whether anyone
else has had the same experience. If we suddenly awake
imagining that we have heard a banging at the front
door, and if there is somebody else in the room who
shows no sign of having heard anything, we conclude
at once that we were dreaming. But, if somebody else
also heard the noise, we have no further doubt that ours
was a real sensation.

In the same way, but less frequently, people are sometimes subject to hallucinations when awake. If somebody tells us that he has seen a transparent old gentleman
clanking chains about the passages and carrying his
head under his arm, our disposition to believe that he
has seen a ghost will doubtless depend largely upon our
attitude to the general question of the existence of
ghosts. But, whatever that attitude may be, our belief
would be enormously strengthened if we found that
other people present at the same time and place had

experienced the same sensations. In fact, a very little reflection will show that our recognition of the possibility of dreams and hallucinations is based almost entirely on the fact that there are circumstances in which the sensations of one person may not be shared by others ; dreams and hallucinations are simply mental experiences which, though almost indistinguishable by the percipient from sensations, are distinguished from sensations by being peculiar to the percipient and in not being shared by others. The community of sensations is our chief and final test that experiences are true sensations such as give information about the external world ; if we apply other tests, it is only because this chief test is not available, and any other tests we may apply are based on the results which we are accustomed to obtain with this chief test.

OUR BELIEF IN " OTHER PEOPLE "

But now we must inquire a little more deeply and face a difficulty. We believe in the external world because the sensations of other people agree with our own. But what reason have we to believe that there are other people ? In our discussion hitherto we have spoken of the world as divided into two parts, man and nature, and we have regarded the external world as the same thing as nature. But it is not really the same thing. If I divide the world into man and nature, you are not part of nature ; but if I divide the world into an external and an internal part, you are part of the external part. " You " are not " me " and " I " am not " you," you are part of my external world and I am part of yours. Nature, the part of the external world that is not man, is the same thing as that part of the world which is external to *all* men ; but it is not the same thing as my external world or as your external world. Accordingly, if I am asking what evidence there is for an external world, I must first make up my mind whether you and other people are to be regarded as part of it.   If I do not regard you as part of the external world, it is unreasonable to regard the community of your sensations with mine as giving me evidence of an external world.   For that community only gives me such evidence if you are external to me ; if your sensations are internal to me, it is clear that the fact that they agree with mine does not justify the argument for the external world that we have just been considering.   On the other hand, if I know that you are part of the external world, it is quite unnecessary to examine your sensations and to inquire whether they agree with my own in order to prove that there is an external world ; for if there is not an external world, you cannot be part of it.   It seems that whichever alternative I adopt, the argument for an external world based on the community of your sensations and mine breaks down.   Either I must know already what the argument professes to prove, or it provides no proofs. Let us therefore examine rather more closely why we do all actually believe that there are other people.

Our reason for believing that there are other people appears to be of this kind.   There is attached to me a portion of the external world that I call my body.   It is part of the external world because I can perceive it by my senses ; I can see my own hand, just as I can see any other external object ; I can hear my own voice ; and with my hand I can feel my own eye.   On the other hand, I regard it is peculiarly attached to me, and as "my" body, because it is very intimately under the control of my will.   I can move my hand and I can close or open my eyes by simply desiring to so do ; it is much less affected by obstinacy in the face of my desires than the remainder of the external world.   Now I know that certain changes in this external object which I call my body, changes which I can perceive through my senses, are intimately connected with certain purely internal feelings. Thus, if I bring my hand too near a hot body, I can see that it is snatched suddenly away; and I know that this sudden motion is accompanied by the purely internal feeling of pain and also by certain muscular feelings which are associated with movement of my body. Now I perceive through my senses other parts of the external world which appear very similar to my body, and these objects undergo associated changes very similar to those which take place in my body. Thus I may see another object, very like my hand, approach the same hot body; and if I see that, I shall see it snatched away again, exactly as I see my hand snatched away. But this time I shall not experience any feeling of heat or any feeling of muscular motion.

To explain these observations I imagine that, just as there is intimately associated with my body a mind, namely, my own mind, so there is intimately associated with each of these other objects, so similar in appearance and in behaviour, another mind; I call these other objects " other persons' bodies," and the minds which I imagine to be associated with them I call " other persons' minds " or simply " other persons." I believe there are other people because I see other bodies reacting in the same way as my body; and, if any reaction of my body is accompanied by some event in my mind, I suppose that the reactions of these other bodies are accompanied by similar events in the minds of the other people.

I do not propose to inquire whether this line of argument is justified (if anything so elementary and so fundamental to all thought can be called argument) or whether it avoids the difficulty to which attention has been called. The reader must inquire for himself whether he can put the evidence for the existence of other people in a form which is wholly convincing and is also such that it is possible to base on the existence of other people an argument for and a criterion of the external world, without lapsing into the fallacy of a circular argument which assumes what it pretends to prove. As we shall see in a moment, it is not relevant to our inquiry to decide whether such arguments are justified or, indeed, whether it is possible to produce any valid arguments for the existence of other people and of the external world. All that I am concerned with here is to draw attention to the ideas which undoubtedly underlie our habitual and common-sense distinction between the internal and the external world, or between other people and ourselves, on the one hand, and nature on the other. The ideas which are important for our further inquiry are :

1. That the conception of " myself," on which is founded the conception of all other people, is intimately connected with the mental experiences which we call will or volition. A person is something that wills ; volition is the test of personality ; nothing is a person or has personality (at least of the human type) unless it is characterized by a will ; all exercise of will is inseparable from the recognition of a person who exercises it ; and everything that is directly subject to the same will is part of the same person.

2. Our belief in the external world, or at least of that part of it which is called " nature," is based on our perceptions received through our sense organs. And we believe that these perceptions inform us of the external world, partly because they are independent of our wills, but more because other people agree with us in those sensations.

3. Our belief in other people is based on an analogy between the behaviour of their bodies and the behaviour of our own. If the actions of other bodies are similar to those of our own, and if those actions in our bodies are accompanied by certain thoughts in our minds, then we

believe that there are similar thoughts in the minds of the other persons whose bodies behave similarly to our own.

A DEFINITION OF SCIENCE

This discussion was started by the suggestion that we could answer our question, What is Science ? by saying that science consists in the study of the external world of nature. For reasons which have been given already, and for others which will appear in due course, I propose to reject that definition of science. In its place I propose to put another, which could have been offered before, but, if it had been offered before the discussion which has just ended, it would hardly have been intelligible. This definition is : Science is the study of those judgments concerning which universal agreement can be obtained.

The connexion between this definition and the ideas that we have been considering is obvious. It is the fact that there are things concerning which universal agreement can be obtained which gives rise to our belief in an external world, and it is the judgments which are universally agreed upon which are held to give us information about that world. According to the definition proposed, the things which science studies are very closely allied to those which make up the external world of nature. Indeed, it may seem at first sight, that we are practically reverting to the definition of science as the study of nature and that there is little difference except in words between the definition which is proposed and that which has been rejected.

But there are two very important differences. In the first place the mere omission of such terms as " nature " and the " external world " is important. For these terms represent inferences from the judgments that we are considering. Nature is not sensations or judgments concerning which there is agreement ; it is something which we infer from such sensations and judgments. And this inference may be wrong. As has been said, nobody maintains that it is entirely wrong, but it is very strongly held in some quarters that some parts of the inference usually made by common sense are wrong and seriously misleading. If we call science the study of nature, we are bound to admit that, if the common-sense view about nature is largely mistaken, there must also be a considerable element of doubt as to the real value of the conclusions of science itself ; in other words, science must to some extent be subordinated to philosophy, in whose province lies the business of deciding the worth of the popular conception of nature. Against such subordination students of science have always protested : and they can maintain their protest if they adopt the view that science studies, not the external world, but merely those judgments on which common sense, rightly or wrongly, bases its belief in an external world. And it may here be noted that among the difficulties that are avoided by the definition are those to which reference was made on p. 23.

But there is a much more important difference. It is true that the popular belief in the external world is founded primarily upon the fact of agreement about sensations ; but, in deciding what part of our experience is to be referred to that external world, common sense does not adhere at all strictly to the criterion on which that belief is ultimately based. We do not ordinarily refuse to regard as part of the external world everything about which there is not universal agreement. A very simple example will illustrate this point. A moment ago a book fell from my table to the floor : I heard a sound and, looking round, saw the book on the floor. Now I had no hesitation in referring that experience to something happening in the external world ; but there was not, and there cannot be, universal agreement about it or indeed any agreement at all. For I am alone in the room and nobody but myself has ever had, or can ever have, any share in that experience. Accordingly our definition of science excludes that experience of mine from the judgments which science studies, although to common sense it was certainly an event in the external world.

Such a simple example indicates at once how very much stricter than the common-sense criterion of externality is the criterion which must be satisfied before any experience is admitted by our definition as part of the subject-matter which science studies. Science, as we shall see, really does maintain the criterion strictly, while common sense is always interpreting it very loosely. I do not mean to assert that common sense is wrong to apply a less strict criterion—that is a question which lies far outside our province ; all that I mean is that any experience which fails to satisfy the strict criterion of universal agreement, though it may be quite as valuable as experience which does satisfy it, does not form part of the subject-matter of science, as we are considering it. Here is the distinction between modern science and the vaguer forms of primitive learning out of which it grew. When the possibility of applying the strict criterion of universal agreement was realized, then, for the first time in the history of thought, science became truly scientific and separated itself from other studies. All the early struggles of science for separate recognition, Bacon's revolt against mediæval learning and the nineteenth-century struggle of the " rationalists " against the domination of orthodox theology, can be interpreted, as we shall see, as a demand for the acceptance of the strictly applied criterion of universal agreement as the basis for one of the branches of pure learning.

IS THERE UNIVERSAL AGREEMENT ?

But objections are probably crowding in upon the reader's mind. The more he thinks about the matter, the more impossible it will appear to him that truly and perfectly universal agreement can be obtained about anything. The scientific criterion, he will think, may be an ideal, but surely even the purest and most abstract science cannot really live up to it in a world of human fallibility. Let us consider for the moment some of the objections that will probably occur to him.

In the first place, he may say that it is notorious that men of science differ among themselves, that they accuse each other of being wrong, and that their discussions are quite as acrimonious as those of their philosophical or linguistic colleagues. This is quite true, but the answer is simple. I do not say that all the propositions of science are universally accepted—nothing is further from my meaning ; what I say is that the judgments which science studies and on which its final propositions are based are universally accepted. Difference of opinion enters, not with the subject-matter, but with the conclusions that are based on them.

In the second place, he may say that, if absolutely universal agreement is necessary for the subject-matter of science, a single cantankerous person who chose, out of mere perversity, to deny what every one else accepted could overthrow with one stroke the whole fabric of science ; agreement would cease to be universal ! Now this objection raises an important issue. How do we judge what other people think, and how do we know whether they do agree ? We have already discussed this matter from the standpoint of common sense and stated our conclusion on p. 25. But, here again, science, though applying generally the same criterion as common sense, insists on a much stricter and deeper application of it. We judge men's thoughts by their actions. In common life we generally use for this purpose one particular form of action, namely, speech : if a man says " I see a table " I conclude that the thoughts in his mind are the same as those in my mind when I say " I see a table." And men are generally so truthful that we do not often need to examine. further. But sometimes we may suspect that a man is wilfully lying and that the relation between his words and thoughts is not normal (although it is again a relation of which we have some experience in our own minds), and we can often detect the lie by examining other actions of his. Thus, if he says that he cannot see a table, we may not be able to make him change his assertion ; but we may be able to induce him to walk across the room, after having distracted his attention from the matter, and then note that he, like ourselves, walks round the table and does not try to walk through it. Such tricks are familiar enough in attempts to detect malingerers in medical examinations. But what I want to point out here is that the method can only be applied to detect lies about a certain class of matters. If a man says that he does not believe that 2 and 2 make 4, or holds that an object can be both round and square, I do not see that we have any way whatsoever to prove that he does not believe what he says he believes. And the distinction is clear between the matters in which lying and imposture can be detected and those in which it can not. As we detect imposture by examining a man's actions, it is only in thoughts and beliefs that affect his actions that we can find out certainly what he thinks or believes. There may be actually universal agreement on the proposition that 2 and 2 make 4, but in this case the objection that we are considering is valid. A single denier could upset that universal agreement, and we should have no way of discounting his assertion and proving that the agreement really is universal. Accordingly, in defining science as the study of judgments concerning which universal agreement can be obtained, we are limiting science to judgments which affect action and deliberately excluding matters which, though they may actually be the subject of universal agreement, do not affect action. This conclusion is important, because it enables us to separate science clearly from pure mathematics and logic ; but space cannot be spared to pursue this line of thought beyond a bare reference to it.

A man may also fail to join in the general agreement, not because he is lying, because he is suffering from some hallucination. This possibility was noticed before (p. 22), and then we distinguished hallucinations from true sensations by the fact of the agreement of others. But now we are applying the test of agreement much more strictly, and the mere fact that the man is under an hallucination and does not agree with others is sufficient to make the test fail. However, the difficulty can be overcome in exactly the same way as that arising from lying. We study *all* the man's actions, and we usually find that, while some of them are consistent with his assertion that he does not agree, others are inconsistent with that assertion ; and those that are inconsistent are those which we know, from our own internal experience, to be less directly connected with consciousness and less liable to aberration. Our test, once more, is always whether the man acts on the whole as we should act if we shared the thoughts which he professes. Curious instances of this nature have occurred in actual science ; there have been people who professed to be able to see or to hear or to feel things which other men could not see or hear or feel. But so far the difficulty has always been removed by setting " traps," even if the honesty of the man is beyond doubt, and showing that his actions in general are not consistent with his professions.

But I mention this matter for another reason. There are persons under what we may call permanent hallucinations : colour-blind people are an instance. There are people who say that, to them, two objects, which to normal people appear one as pink and the other as greenish-blue, appear exactly the same colour ; and no traps set for them will show any inconsistency in their judgment. They will maintain their position when all their interests lie in an ability to distinguish the colours. In such cases universal agreement cannot be obtained. Are the judgments to be excluded from the subject-matter of science ? The answer is, Yes ; they are excluded. And the fact that they are excluded is a support for the definition of science which has been offered ; for there is no doubt that they would have to be included if science studied simply the properties of the external world. Strange as it may appear to the uninitiated, colour, judged by simple inspection, is not a scientific conception at all, and it is not a scientific conception because universal agreement cannot be obtained about it. The procedure adopted is this. We find that normal people regard the objects A, B, C, . . . as all pink and the objects X, Y, Z, . . . as all blue ; colour-blind people, on the other hand regard A, B, C, . . . X, Y, Z, as indistinguishable in colour. But we find also that there is some other property in which both normal and abnormal people find that A, B, C, . . . agree and that X, Y, Z, . . . agree, while in respect of this property both normal and abnormal find that A, B, C, . . . differ from X, Y, Z, . . . When we find that, we regard this new property as the true and scientific test of colour ; for about this property we can obtain universal agreement. And we call some people abnormal, not merely because they fail to agree with the majority, but because they fail to make a distinction where it is universally agreed that there is a distinction.

To make this important matter clear, it may be well to suggest a procedure which might be adopted. We might make both the normal and the abnormal people look at the objects through a red glass. Through the glass, of course, everything will look the same colour to both normal and abnormal, but different objects will appear different shades of the same colour. The pink objects A, B, C, . . . will all appear the same shade, and so will the greenish objects X, Y, Z . . . ; but the former will appear a lighter shade than the latter ; and they will appear a lighter shade, not only to the normal people, who see the difference of colour when the red glass is not interposed, but also to the abnormal,who do not see this difference. Here then universal agreement has been attained ; every one agrees that through the red glass the objects look different. Accordingly we regard the appearance through the red glass as a better basis for science than the appearance without the red glass ; we say that scientifically the objects are different in colour if they appear different through the red glass, and we call one set of people normal and the other abnormal because one set agree and the other do not with the distinction based on this truly scientific criterion.

It is a very remarkable fact that, wherever we find such abnormal people under permanent hallucinations (and we find them in regard to all the senses), we can always find another test which, in the manner just described, enables us to restore universal agreement. It is this fact, which could not be anticipated, which makes science possible, and gives its great importance to the test of universal agreement.

But perhaps the reader may doubt whether there are really judgments about which *every one* agrees, if we are allowed to include people with the most extremely abnormal sensations, such as the totally blind or the totally deaf. The doubt can only be removed by quoting an example which can easily be given. Every one who has any sense-perceptions at all, and can come into any contact with the external world, has the feeling that events occur at different times and that some occur *before* others. This is an example of a judgment concerning which there appears to be the truest and most perfect universal agreement. If one person A judges that an event x occurs before an event y, then anyone else, however abnormal his sensations so long as he can experience the events at all, will also judge that x occurs before y; he will never judge that x occurs after y. If the reader considers this example, I think he will feel that in such a judgment of the order in which events occur it is almost inconceivable that there should be anything but the most perfect universal agreement. Such judgment, and such only, form the proper basis of science.

However our objector may make a last stand. He may say that, though it is barely conceivable that there should be disagreement about such a matter, barely conceivable events do sometimes occur. It is just possible that disagreement might arise where now there is the most perfect agreement; what would science do then? The question is unanswerable. It is quite impossible to say what we should do if the world was utterly different from what it is; and it would be utterly different from what it is if there were not judgments concerning which universal agreement is obtainable. It would be a world in which there was no "external world." For though, as has been urged, the general agreement on which popular ideas about the external world is based is not always as perfectly universal as is demanded by the criterion set up by science, a deeper inquiry than we can under take here would show that common sense, just as much as science, does employ conceptions which would be meaningless if, in the last resort and in some cases, perfectly universal agreement

were not obtainable.   That is the true answer to all the
objections that we have just been considering ;   it has
been useful to consider them, because we have been
enabled thereby to bring to light some matters important
in the procedure of science ; but the answer to all objections based on the difficulty that might conceivably
be encountered in obtaining universal agreement is that
such agreement is actually obtained, and that all our
practical life and all our thought are based on the
admission that, in some matters but not in all, it is
actually obtained.

There is, however, an objection of another kind which
may yet be raised, but, since the discussion of it leads
us directly into more strictly scientific inquiry, it will be
well to leave it to open a new chapter.

CHAPTER III

THE LAWS OF SCIENCE

WHY DOES SCIENCE STUDY LAWS ?

THERE was quoted on p. 28 an example of an experience which could not be the subject-matter of science, according to our definition, because there could not be universal agreement about it. A book fell on the floor when only one person was in a position to observe the fall. Now it may be urged that this example is typical, not of a small and peculiar class of events occurring in the external world of nature and perceived by the senses, but of all such events. No event whatever has been observed by more than a very small minority of mankind, even if we include only persons who are all alive at the same time ; if we include—and our definition suggests that we ought to include—all men, past, present, and future, it is still more obvious that there can be no event concerning which they can all agree ; for there is no event which they can all perceive. Are we then to take the view that no event whatever is the proper subject-matter for science ? And, if we take that view, what is there left in the external world which can properly be such subject-matter ?

The answer is that we *are* to exclude every particular event from the subject-matter of science. It is here that science is distinguished from history ; history studies particular events, but science does not. What then does does science study ? Science studies certain relations between particular events. It may be possible for every one to observe two events each of a particular kind, and to judge that there is some relation between those events ; although the particular events of that kind which they

37

observe are different. Thus, in our example, it is impossible for every one to observe that a particular book fell to the floor and made a noise on striking it ; but it is possible for every one to observe that, *if* a book is pushed over the edge of the table, it *will* fall to the floor and *will* make a noise on striking it. Concerning that judgment there can be universal agreement ; and that agreement will not be upset, even if somebody has never actually observed a book fall ; so long as he agrees when at last he is placed in the necessary circumstances, that a book will fall and that, when it falls, it will make a noise, then universal agreement is secured.

If we could imagine ourselves without any experience of the external world derived from our senses, we might doubt whether there actually are such relations concerning which universal agreement can be obtained ; we might expect that it would be as impossible to find universal relations between events as to find universal events. But we all know from our experience that there are such relations and we know of what kind these relations are. They are of the kind that have just been indicated ; the universal relations that we can state are between events which are such that, *if* one event happens, *then* another event happens. Again, there might conceivably be other relations between events of a different kind, yet of the same universality ; actually there are not—at any rate if we interpret the relation just stated correctly. There is a certain class of relations between events for which universal agreement can be obtained, which is thereby distinguished from other classes for which it cannot be obtained. Indeed, we might almost say that it is only this class which can be the subject of universal agreement ; for the necessity that all men, even if they live at different times, should agree imposes limitations on the form of the relation. But we need not inquire into this abstruse matter ; all that is necessary for our purpose is to recognize that there are certain relations between events concerning which all men can agree.

Our definition then limits science to the study of these special relations between events. And this conclusion, though the form in which it has been expressed and the reasons alleged for it may be unfamiliar, is very well known and widely recognized. For the relation of which we have spoken is often called that of " cause and effect " ; to say that, if a book falls off the table, it will make a noise when it strikes the floor is much the same as to say that the noise is the effect of the fall, and the fall the cause of the noise. Again, assertions of cause and effect in nature are often called " laws " or " laws of nature " ; in fact, the assertion that a book, or any other object, will fall if unsupported is one of the most familiar instances that is often offered of one of the most widely known laws, namely the law of gravitation. Accordingly, all that we have said is that science studies cause and effect and that it studies the laws of nature ; nothing can be more trite than such a statement of the objects of science. Indeed, I expect that some readers thought that a great deal of unnecessary fuss was made in the previous chapter and that all our difficulties about the relation between science and nature would have vanished, if it had been said simply that science studied, not nature, but the laws of nature.

However here, as so often, the popular view, though it contains a large measure of truth, is not the whole truth. The meaning popularly attached to " cause and effect " and to " laws " is too loose and vague ; " cause and effect," in the conversational sense, includes some relations which are not studied by science and excludes some that are ; the assertions which are popularly regarded as laws are not invariably scientific laws, and there are many scientific laws which are not popularly termed so. The value of our definition is that it will enable us to give a more precise meaning to these terms, and to show clearly why and where scientific and popular usage differ. Accordingly, in the rest of this chapter we shall examine the matter more closely.

First, we may note that there is an apparent difference between the popular conception of the part played by laws in science and that laid down by our definition. It is probably usually thought that it is the aim and object of science to discover laws, that laws are its final result. But according to our view nothing can be admitted to the domain of science at all unless it is a law, for it is only the relations expressed by laws that are capable of universal agreement. Laws are the raw material, not the final product. There is nothing inconsistent in these two statements, but the mode in which they are to be reconciled is important. Laws are both the raw material and the finished product. Science begins from laws, and on them bases other laws.

To understand how this may be let us take an example of a law ; that used already is not very suitable for the purpose ; the following will serve better : A steel object will rust if exposed to damp air. This is a law ; it states that if one event happens another will follow ; although it is the result of common observation, it would usually be regarded as lying definitely within the province of science. But now let us ask what we mean by a steel object, or by " steel." We may say that steel is a hard, shining, white substance, the hardness of which can be altered by suitable tempering, and which is attracted by a magnet. But, if we express what we mean by steel in this way, we are in effect asserting another law. We are saying that there is a substance which is *both* shining, white, *and* capable of being tempered, *and* attracted by a magnet ; and that, *if* it is found to be white and capable of tempering, *then* it will be magnetic. The very idea of "steel" implies that these properties are invariably associated, and it is just these invariable associations, whether of "properties" or of "events," that are expressed by laws. In the same way "rust" implies another set of associated properties and another law ; rust would not be rust, unless a certain colour were associated with a powdery form and insolubility in water.

And we can proceed further and apply the same analysis to the ideas that were employed in stating that the properties of steel are invariably associated, or, in other words, that there is such a thing as steel. For instance, we spoke of a magnet. When we say that a body is a magnet, we are again asserting an invariable association of properties ; the body will deflect a compass needle and it will generate an electric current in a coil of wire rotated rapidly in its neighbourhood. The statement that there are magnets is a law asserting that these properties are invariably associated. And so we could go on finding that the things between which laws assert invariable relations are themselves characterized by other invariably associated properties.

This, then, is one of the ways in which laws may be both the original subject-matter and the final result of science. We find that certain events or certain properties, A and B, are invariably associated ; the fact that they are so associated enables us to define a kind of object, or a kind of event, which may be the proper subject-matter of science. If the object or the event consisted of A and B without any invariable association between them, it might be a particular object or a particular event, and might form an important part of the popular conception of the external world, but it would not be proper subject-matter for science. Thus the man Napoleon and the battle of Waterloo are an object and an event consisting of various properties and events ; but these properties and events are not invariably associated. We cannot, by placing ourselves in such circumstances that we observe some of the properties (for instance short stature, black hair, and a sallow complexion), make sure that we shall observe the other properties of Napoleon. On the other hand, iron is a kind of object suitable for the contemplation of science, and not a particular object, because, if we place ourselves in a position to observe some of the properties of iron, we can always observe the other properties. Now, having found an A and B invariably associated in this way, and therefore defining a kind of object, we seek another set of associated properties, C and D, which are again connected by a law, and form another kind of object. We now discover that the kind of object which consists of A and B invariably associated is again invariably associated with the kind of object which consists of C and D invariably associated ; we can then state a new law, asserting this invariable association of (AB) with (CD) ; and this law marks a definite step forward in science.

If it is in such a way that science builds up new laws from old it clearly becomes of great importance to decide what are the most elementary laws on which all the others are built. It is obvious that the analysis which we have been noticing cannot be pushed backwards indefinitely. We can show that in a law connecting X with Y, X is the expression of a law between A and B, and Y of a law between C and D ; we may possibly be able to show again that A is the expression of yet another law connecting some other terms, a and b. But, in the last resort, we must come to terms, a and b, which are not resolvable into other laws, and which, therefore, are not proper subject-matter for science by themselves, but only when they occur in the invariable association (ab). What, then, are the terms at which we arrive at length by this analysis ? What are the irresolvable laws which must lie at the basis of all science ?

No more difficult question could be asked, and I cannot pretend to answer it completely, even for that small branch of science which is my special study. The reason for the difficulty is interesting, and we must examine it.

Let us return to our first " law," namely that steel will rust if exposed to damp air.   I said that the use of the word steel implied an invariable association of properties which is asserted by the more elementary law : There is such a thing as steel.   But if we look at the matter closely we shall see that this is not really a law. For there are many kinds of steel ; the substances, all of which the man in the street would call equally " steel," are divided by the fitter into mild steel, tool steel, high-speed steel, and so on.   And the scientific metallurgist would go further than the fitter in sub-division ; he would recognize many varieties of tool steel, with slightly different chemical compositions and subjected to slightly different heat-treatments, which might be all very much the same thing for the purposes of the fitter.   But if we say that there are several kinds of steel, we are in effect saying that the association of the properties of steel is not invariable, that there can be many substances which, though they agree in some of their properties, differ in others.   Thus everything anybody would call steel contains, according to the chemist, two elements, iron and carbon ; but most steel contains some other element as well as these two, and these other elements differ from one steel to another ; one contains manganese, another tungsten, and so on.   It is not a law that every substance which contains iron and carbon (and has certain physical properties of steel) contains manganese ; for there are substances which agree in all these respects, but differ in containing nickel in place of manganese, and in certain other physical properties which are not common to all steels.

There may seem to be an easy way out of the difficulty

raised by finding that there is really no such thing as
" steel." It has been implied that there are certain
properties common to all steels. If we make the word
steel mean anything which has these common properties,
whatever other properties it may have, then, since these
properties are invariably associated, the proposition that
there is steel (in this sense) will be a true law. But if
we examine the matter closely enough, we find that there
are not really any properties common to all steels ; we
can find common properties only if we overlook distinctions which are among the most important in science.
All steels, we may say, contain iron and carbon and all
are capable of being tempered. But they do not all contain the same amount of iron and carbon, nor are their
tempering properties all the same ; and the variation
in the amount of carbon they contain is associated with
important variations of their tempering properties. As
we shall see—if, indeed, it is not obvious—one of the
most important distinctions which science makes is
between objects or substances which have all the same
property, but have it in different degrees ; the study of
such distinctions is measurement, and measurement is
essential to science. The deeper we inquire, the less we
shall feel inclined to regard the statement, there is
steel, as a law, asserting invariable associations We
shall want to break this law up into many laws, one
corresponding to each of the different kinds of steel that
we can recognize by the most delicate investigation ;
when we have pushed these distinctions to the utmost
limit, then, and not till then, we shall have arrived at
laws stating truly invariable associations between the
various properties of these different kinds of steel.

Here we meet with a process in the development of
science precisely contrary to that which we considered
before. We were then considering the process by which
science, starting from a relatively small number of laws,

found relations between the objects of which they are the laws, and so arrived at new and more complex laws. In the second process, science takes these simple laws. analyses them, shows that they are not truly laws, and divides them up into a multitude of yet simpler laws. These two processes have been going on concurrently throughout the history of science ; in one science at one time one of the processes will be predominant ; in another science at another time, the other. But on the whole the first process is the earlier in time. Science started, as we have seen, from the ordinary everyday knowledge of common sense. Common sense recognizes kinds of objects and kinds of events, distinguished from particular objects and particular events by the feature we have just discussed ; they imply the assertion of a law. Thus all " substances," iron, rust, water, air, wood, leather, and so on, are such kinds of objects ; so again are the various kinds of animals, horses, sparrows, flies, and so on. Similarly common sense recognizes kinds of events, thunder and wind, life and death, melting and freezing, and so on ; all such general terms imply some invariable association and thus are, if the association is truly invariable, proper matter for the study of science. And science in its earlier stages assumed that the association was invariable, and on that assumption proceeded to build up laws by the first process. It found that *iron* in damp *air* produced *rust* ; that *poison* would cause *death*.

But, as soon as this process was well under way, the second process of analysis began ; it was found that the association stated by the laws implied by the recognition of such objects was not truly invariable. This discovery was a direct consequence of the first process. Thus, until we have discovered that steel in general rusts, we are not in a position to notice that there are some steels which do not rust. When we have found that there are certain substances, otherwise like steel, but differing from other steels by being rustless, we are for the first time in a position to divide steels into two classes, those which do and those which do not rust, and so to analyse the single law implied by the use of the term " steel " into two laws, one implied by the term rusting steel, and the other by the term rustless steel. And so, again, when we have found that steel is attracted by a magnet, we are first in the position to notice that different objects, hitherto all called magnets, differ somewhat in their power of attracting steel ; we can break up the single law, There are magnets, into a whole series of laws asserting the properties of all the various magnets which are distinguished by their different power of attracting steel.

This is actually the history of scientific development, so far as the discovery of laws is concerned. And now we can see why it is so difficult to say what are the fundamental and irresolvable laws on which science is ultimately built. Science is always assuming, for the time being, that certain laws are irresolvable ; the law of steel, for example, in the early stages of chemistry. But later it resolves these laws, and uses for the purpose of the resolution laws which have been discovered on the assumption that they are irresolvable. At no stage is it definitely and finally asserted that the limits of analysis have been reached ; it is not asserted even in the most advanced sciences of to-day ; it is always recognized that a law which at present appears complete may later be shown to state an association which is not truly invariable. Moreover, the intermingling of the two processes leads to the result that a law which is regarded as final in one connexion is not regarded as final in another. We use the law that there is steel to assert the law that there are magnets, and at the same time use the law that there are magnets to assert the law that there is steel !

If we attempted to describe science as a purely logical

study in which propositions are deduced one from the other in a direct line of descent from simple ultimate assumptions to complex final conclusions, this double rôle of laws, partly assumptions and partly conclusions, would cause grave difficulty. All scientific arguments would appear " circular," that is to say, they would assume what they pretend to prove. But the result that follows from our discussion is not that science is fallacious, because it does not adhere to the strict rules of classical logic, but that those rules are not the only means of arriving at important truths. And it is essential to notice this result ; for, since logic was the first branch of pure learning to be reduced to order and to be brought to something like its present position, there has been a tendency in discussions of other branches—and especially in discussions of science—to assume that, if they have any value and if they do really arrive at truth, it can only be because they conform to logical order and can be expressed by logical formulas. The assumption is quite unjustifiable. Science is true, whatever anyone may say ; it has, for certain minds, if not for all, the intellectual value which is the ultimate test of truth. If a study can have this value and yet violate the rules of logic, the conclusion to be drawn is that those rules, and not science, are deficient. Nevertheless, while it is important to insist that science is not necessarily bound by logical formulas, it may be well to point out that the difficulty which we have been noticing can be overcome to some extent. The difficulty arises because we have regarded all the different laws of science as different propositions, some of which give rise to others. It would probably be more accurate to regard all the so-called laws of science as one single law which is always being extended and refined ; and if we take that view there can be no question of deducing one law from another ; the difficulty does not arise. Much might be said in further explanation and extension of this attitude ; but space forbids a more lengthy discussion, and, with this hint, the matter must be left.

It may be noted that in the actual practice of science none of these difficulties and complexities arise. Every science starts, as has been said, from the crude and vague laws which have been elaborated as the result of that continuous tradition of experience which is called common sense. And, just because they are so intimately part of common sense, there is usually no difficulty whatsoever in obtaining for them the universal agreement which makes them the proper subject-matter of science. It is only when science gets to work and, instituting a much deeper and more thorough inquiry than common sense would ever institute, finds that the relations asserted by the laws are not strictly invariable, that the question of doubting the laws arises ; and the very inquiry which suggests the doubts suggests also how the laws may be amended so that once more, for the time being, universal assent for them may be obtained. It is not actually difficult to get people to agree that there are such things as air and water ; the actual difficulty is rather to make them see that what they call air and water are really many different substances, all differing slightly by small distinctions which have been overlooked. When we study the history and development of any actual science —and it must be remembered that this book is only intended to be an introduction to such study—we do not find actually that difficulties are continually raised by a failure to obtain universal agreement ; though at a later stage it is easy to see that the supposed laws of an earlier stage were not true laws and that agreement could not have been obtained for them, at any one stage the distinction between the laws which are assumed as funda-mental and those which are based on them is perfectly clear and definite. The criterion of universal agreement is important because it gives a reason why we do actually select for the study of science those portions of experience which are actually selected ; but it is not the criterion which we consciously apply. The conscious criterion for the subject-matter of science is rather that it has been regarded hitherto as connected together by a relation of invariable association such as is asserted by a law.

DO LAWS STATE CAUSES AND EFFECTS ?

So far we have only considered half of the problem of the laws of science. We have expanded and made more precise the conception of a law of nature, have considered why such laws are of such supreme importance for science, and have inquired how they can be at once its starting point and its goal. A law, we have concluded, is the assertion of an invariable association, and the events or properties or other things that it declares to be invariably associated are themselves collections of other invariably associated things. But we have not attempted to ask further what is meant by " invariable association." We noticed in passing at the outset that it was often thought that laws were concerned characteristically with relations of cause and effect. A cause and its effect are invariably associated. The view is therefore suggested that by invariable association we mean simply the relation of a cause to its effect. Is that what we mean ? This is the other half of our problem and to it the rest of the chapter must be devoted.

We must naturally start by asking ourselves what exactly we mean (or what we should mean) by " cause and effect." This is a matter on which there has been much discussion ; but the idea which underlies most frequently the use of the terms seems to be this. We imagine that whenever an event B happens, it happens only because it has been preceded by some other event

A ; and, on the other hand, if A happens, it is sure to be followed in due course by B. When we can discover such a relation between two events, we say that A is the cause of B and B the effect of A. A single example will suffice for illustration. If my finger bleeds, it is *because* I have cut it. The cutting, which necessarily precedes the bleeding, is the cause ; the bleeding which necessarily follows the cutting is the effect.

However, this simple and familiar notion, like so many equally simple and familiar to a first glance, appears rather more complex and intricate on further examination. The many difficulties which might be and have been raised to the acceptance of this simple view are not strictly relevant to our present purpose, but a few of them may be noted for the information of the reader unaccustomed to philosophical discussion. The first difficulty is that there are certainly pairs of events, A and B, one invariably preceding the other, which we do not regard as cause and effect ; for instance, birth invariably precedes death, and yet we should not accept readily the conclusion that birth is the cause of death. Again, sometimes B, though always following A, also always precedes another A ; day always follows night, but it also always precedes night ; is day or night the cause, or is there no relation of cause and effect involved ? Once more, even when we are clear that there is a relation of cause and effect involved, it is often difficult to say precisely which, out of many alternations, is the cause. Death, for instance, may be the effect of natural causes, or of a hundred forms of accident or violence. We know that it must always be the effect of one of them, but we are so uncertain of which is the cause in each particular case that a special form of inquiry is thought necessary. How is this uncertainty consistent with the invariable sequence of effect after cause which seems assumed by the use of those words ? Such difficulties as these undoubtedly suggest that by cause and effect we mean something rather more abstruse and certainly more obscure than the simple invariable sequence of one event after another, which seems usually to be regarded as constituting the causal relation.

But this is not what we have to consider. For those who have seriously maintained that the business of laws is to state relations of cause and effect have always regarded such relations as consisting merely of invariable sequences. It is possible that if the terms are used in this sense they do not coincide exactly with common usage, but, if that is so, it will only be one more of the innumerable examples where science has diverted a term slightly from its sense in popular discourse. What we have to ask is whether, in discovering scientific laws, we are simply establishing invariable sequences in which one event or set of events follows after another.

It may be admitted without further discussion that some part of the laws of science do actually consist of statements of invariable sequences. So much follows at once from our previous discussion. For, though we have spoken hitherto more vaguely of invariable association rather than of invariable sequence, it is obvious that, if there is such a thing as an invariable sequence, it is *one form* at least of invariable association and possesses the qualities which we concluded were necessary to make a relation the proper subject-matter of science ; invariable sequence is a relation concerning which universal agreement might be obtained, just because it is invariable. On the other hand, it seems certain that there are such things as invariable sequences, for it is doubtless within the province of science to predict future events, for example the motions of the stars and the changes of the weather ; and how could prediction from present to future be possible, unless it were possible to discover sequences of events which are invariable and which always recur ?

But it is much more doubtful whether it is only, or even mainly, such sequences which are studied in the establishment of laws. Indeed, some of the examples of laws that have been quoted already seem to state relations which are not sequences. For instance, we spoke of the law of the association of the properties of steel or of a magnet. But properties are not events which follow each other. It is not necessary, in order to prove that a substance is steel, always to observe that it is attracted by a magnet *before* it is observed that it will rust in damp air ; there is no time-relation of any kind between the two properties. The properties of a single substance, the invariable association of which is asserted by the law of that substance, are something quite independent of the times at which they are observed. They differ completely in this matter from events which are related to each other as cause and effect.

And there are scientific laws of another kind which are not concerned with the invariable sequences that constitute cause and effect, namely, numerical laws, of which we shall have much to say later. Important examples of such laws are those which state that one magnitude is proportional to another. For instance, Ohm's Law states that the electric current through a conductor is proportional to the electrical pressure between its ends, so that if the pressure is doubled, the current is doubled. Here, again, there is no time-relation involved ; the law states something about numbers and the size of the things that they represent ; there is no idea of one thing being before or after another.

But, if there are so many and such important laws which are obviously not concerned with cause and effect, how did the idea ever arise that the establishment of causes and effects was the sole or main purpose of scientific laws ? In respect of the first example which has been quoted, that of laws which state the properties of a substance, the answer is undoubtedly that it has not been recognized sufficiently that such propositions are laws. For they are not usually called laws. But the fact that the name is not applied to them is largely the result of history. As we noted, laws of this type are among the results which science accepts in the first instance from the experience of common sense, although it subsequently refines them and may change them almost beyond recognition. Knowledge is dignified by the imposing name of law only when it has been arrived at by deliberate and conscious investigation, and not when, like Topsy, it simply " growed." But it is more difficult to explain why numerical laws, to which the name " law " is applied characteristically, have not been recognized as providing instances to show that cause and effect is not the only relation with which laws are concerned.

I think the real reason is to be found in a confusion between the method by which knowledge is attained and the content of the knowledge once it is attained. What I mean is this. Suppose we were seeking to discover whether Ohm's Law is true. We shall set up instruments for measuring the current and the pressure, and shall then watch how the current changes when we change the pressure. In making such experiments, what we shall actually observe is that a change in current follows a change in pressure ; we shall *first* make deliberately a change in the pressure and *then* observe a change in the current ; in other words, during the experiment the change in current appears as an *effect* of which the change of pressure is the *cause*. But, though it may be maintained that it is by observing such relations of cause and effect that we discover the truth of Ohm's Law, it is not these relations which are stated by the law. It is the numerical relation, and not the relation in time, that is stated by the law. For, if we change our experimental arrangements a little, we shall be able to alter the relation and interchange cause and effect ; we shall be able first to alter the current intentionally and then to observe a change in the pressure. But, though we have thus turned cause into effect and effect into cause, we shall regard the experiments as proving the truth of the same law, because the numerical relation will be unaltered ; the same current will still be associated with the same pressure. Thus, as was said at the start, the law states a relation which is not that of cause and effect, although it may be established by observing such a relation ; there is a distinction between the meaning of the law and the evidence on which it is asserted.

This distinction appears in experiments of all kinds and is hardly separable from the fundamental idea of an experiment. To make an experiment is practically the same thing as to try what is the effect of some cause, and in making it it is impossible not to think of the cause before thinking of the effect. Thus, to revert to an earlier example, if we are proposing to investigate what action damp air has on steel, in order to make the trial we must be thinking about damp air before we can know what that action is. But when we discover that the steel rusts, we see that the rusting is not the effect of the damp air, in the sense that the presence of the damp necessarily precedes the rusting ; we see that the rusting is going on all the time that the steel is exposed to damp ; the exposure and the rusting are concurrent, not con-secutive.

A confusion between the order in which the processes are present in our minds when we are making experi-ments or observations and the relation which the pro-cesses necessarily bear to each other—this confusion is, I believe, the source of the notion, generally prevalent during the last century, that cause and effect (in the sense of a pair of events occurring in an invariable sequence) is a relation of peculiar significance to scientific laws which are based on experiment. Its significance for such laws is very much less than is generally believed. In fact, it would hardly be too much to say that science seeks to avoid entirely the necessity of recognizing such causal relations, even when it is dealing with events which actually do occur in invariable sequences. Consider, for example, a body falling to the ground. Each position of the body invariably follows those higher up and precedes those lower down. We might describe the motion by saying that each higher position is the cause of the lower positions and that the lower positions are the effects of the higher. But actually we do not adopt such a description. We regard the passage of the body through the whole sequence of positions as a single process which is not to be analysed at all ; it is something which, as a whole, may have a cause (such as the presence of the earth which attracts the body) or an effect (such as the noise finally produced by its impact), but in the process itself cause and effect are not involved. This elimination of the causal relation, and its replacement by the conception of a naturally occurring process, is characteristic of all the more advanced sciences.

But, if the relation which laws establish between events or properties or other things is not that of cause and effect, what is it ? That is a very interesting question, but it is too difficult and needs too much detailed knowledge of science for any attempt to be made here to answer it. I think there are many slightly different relations characteristic of laws ; the differences are important and suggestive ; but they all agree in the feature on which such stress has been laid already. They may all be described as various forms of " invariable association " ; and it is because they are all characterized by this invariability that they are capable of being experienced by everybody, and consequently are capable of that universal assent which makes them the proper subject-matter of science.

Nevertheless, there is one particular form of relation involved in laws which can be distinguished from others, and on which emphasis may be laid once more. This relation is that which characterizes what we have called the law of the properties of a substance, or a kind of system, the law, namely, which asserts that there is such-and-such a substance or such-and-such a kind of system, steel or magnets, for example. These laws, in an elementary and imperfect form, are the earliest laws of science, and they retain their peculiar significance through much of its consequent development. The recognition that there are such laws, and that they are laws just as much as others more generally recognized and called by that name, enables difficulties to be overcome which have troubled some of those who have tried to explain the nature of science. One of these difficulties is connected with the " classificatory " sciences, such as the older zoology and botany, or mineralogy. Such sciences seem at first sight to state no laws at all ; they simply describe the various animals, plants, or minerals, and arrange them in groups according to their resemblances or differences, but do not state about them any laws that are usually recognized as such. But, if such sciences do not state laws, why are they regarded as sciences ? We can answer now : They do state laws—laws of the kind which asserts the properties of a kind of system. In establishing that there is such an animal as a cow and pointing out accurately its differences from a sheep, or in investigating the differences between quartz and rock-salt, zoology or mineralogy is discovering laws, and laws of the kind that are of the most funda‧ mental importance. The " classificatory " sciences differ from other sciences in that they confine themselves to laws of this type and (in so fai as they are completely " classificatory ") do not base on them other laws of other kinds.

To the student who, after reading this little book, proceeds to the detailed study of one or more actual sciences, the interesting problem is proposed of distinguishing between the different kinds of laws that are characteristic of them ; for every science has its own peculiarities in the features of its law. But we cannot spend space on this inquiry ; we must now face a different and more pressing problem.

CHAPTER IV

THE DISCOVERY OF LAWS

STATEMENT OF THE PROBLEM

FOR now, having decided what laws are and what they state, we have to ask how they are discovered. Laws state invariable associations; but how can we ever be sure that an association is invariable ? We may have observed an association many times, and have always found that if one of the associated events or properties occurs, the other occurs also ; but if the association is truly invariable, we must know, not only that the association always has been found in the past, but also that it always will be found in the future. Moreover, even if we have found the association every time ᴀn the past that we have looked for it, we clearly cannot know that it has occurred when we have not looked for it. The establishment that an association is invariable and the assertion by a law that it is invariable clearly require that we should be able to judge from the observation of one or several occurrences of it all the other occurrences that may happen or have happened. How can we possibly attain such knowledge ?

One answer to this question is simply that we do not *know*. We can never be certain that an event will happen in the same way that we are certain that it has happened. Indeed, there is a difference in the sense of the word " know " applied to the two cases—a difference in sense which is reflected by the use of different words in most languages. When I have actually experienced an event I have a direct and immediate perception of it which is different in kind, and not merely in degree, from my belief, however confident, that it will happen ;

it is not merely that I have more knowledge of it, but that the knowledge is of a different kind. It is utterly impossible that I should have of the one event the kind of knowledge which I have of the other. If we are to discuss profitably the problem before us, we must remember this difference. We must not seek of events which have not happened, the kind of knowledge applicable only to those which have happened. And again, we must not seek the kind of knowledge—it is once more a different kind—that we have of purely logical or internal propositions. When I say that a black cat is black, I am quite certain that the statement is true because by " a black cat " I *mean* a cat that is black ; to say that a black cat is not black is not untrue ; it is meaningless. The knowledge that I have of the truth of the statement is necessarily different from that which I can have of the statement that there is such a thing as a black cat or that all cats are black ; and the difference is once more in the kind of knowledge and arises from a difference in the kind of statement ; it is not a difference in degree of certainty.

The problem would be expressed better if we merely compared our knowledge of various future events and asked why we are more certain that some will happen than that others happen and how we arrive at this superior knowledge, for then we are sure of comparing always knowledge of the same kind. Of some future events we are as certain as we can be in respect of knowledge of this kind ; we are as certain as we can be that the sun will rise to-morrow. It would be ridiculous to say that we are not certain because we do not feel towards that prediction the same mental attitude that we feel towards the assertion that the sun rose to-day or the assertion that to-day is not to-morrow. For, once more, the difference in mental attitude necessarily arises from the difference in the nature of the statements. All that we can ask relevantly is why we are as certain that the sun will rise to-morrow as we can be of any future event and why we are so much less certain that it will not rain to-morrow.

It is obvious that our certainty in one case and our uncertainty in the other are derived from our previous experience of the happening of similar events, and that the difference in knowledge is due to a difference in that previous experience. Of course this statement does not help us to solve our problem, for since laws are undoubtedly derived from previous experience, it is clear that it is there that the foundation and evidence for them must be found. But the form in which the problem has been put enables us to avoid altogether a question to which those who have discussed the matter have usually devoted most of their attention. They have asked how it is that previous experience gives any knowledge of future experience and what justification there can be for asserting in any case whatever that we have such knowledge. The point of view that I tried to suggest in the last few paragraphs is that this question is essentially unanswerable because it is based on the neglect of the fundamental distinction between different kinds of knowledge. Our " knowledge " of future events simply *is* something based on our knowledge of past events ; when we say that we know something about the future we only mean that we have a mental attitude based on past experience ; and it is absurd to ask why it is based on past experience, for, if it were not so based, it would be something quite different. In my opinion (though the reader should be warned that others would dissent strongly) it can only lead to confusion of thought to attempt to compare this knowledge with other kinds of knowledge and to ask how they stand in relative certainty. And yet some comparison of knowledge of future events with other kinds of knowledge is always intended when it is asked how we have such knowledge from experience of the past.

And because such a question is meaningless the answers given to it are meaningless also. They always consist in some attempt to prove from very abstract and obscure premisses a doctrine called by the high-sounding title of Uniformity of Nature ; it is argued that, for some reason to be found in transcendental philosophy, nature must be such that what is true of her in one part, in one region of space or at one period of time, must be true of her in any other part. But the value of such a doctrine depends entirely on the meaning attributed to " nature." If the world means merely the non-human, external world of common sense (as in Chapter II), then the doctrine is simply untrue. Nature, in this sense, is not uniform ; there are events which happen once and never happen again ; and it is precisely because there are such events that we distinguish between past and future. If it were really true that " history repeats itself," there would be no history ; history is the record of events which have not repeated themselves and the proverb—like almost all proverbs—merely represents an attempt to obtain, by an epigrammatic form, credence for an assertion which nobody would otherwise believe. It is true that many events which do not so repeat themselves, and perhaps the most important of these events, are characteristically human and do not, therefore, form part of common-sense " nature " ; but there are enough non-recurrent events, which have nothing to do with man, to distinguish between past and future and thus to controvert the assertion that all nature is uniform in all its parts.

If, on the other hand, we mean by " nature " in this connexion the carefully scrutinized nature of science, then the doctrine merely states that nature is nature. For this " nature " or external world of science is characterized by and distinguished from everything else by the fact that it is uniform ; for, as we have seen, it is made up of invariable associations concerning which universal agreement can be obtained.   Any part of experience that is not uniform would not consist of invariable associations and would be at once excluded from this closely regulated nature.   Indeed the problem before us is simply that of how we distinguish the uniform from the non-uniform parts of the nature of common sense, for that is our task in establishing the relations which are asserted by laws. To attempt to base a method of making the distinction on the assumption that all nature is uniform is simply to misunderstand the problem that is to be solved.

AN ATTEMPTED SOLUTION

After this clearing of the ground, we can attack the problem.   What is the feature of our previous experience which makes us so certain that the " law " of the rising and the setting of the sun asserts a truly invariable association and, consequently, that the sun will rise tomorrow ?   In answer every one would say that our belief is certain because we have observed the association an immense number of times without observing any failure. And doubtless this is the reason in this particular case, but other instances suggest that the answer is not fundamental or complete.   For there are instances in which an association which has been found as invariable has at length been broken ; and there are instances in which a law is asserted confidently as the result of a single observation, so that there has been no chance of proving any invariability.   The instance of the first kind that is always quoted in these discussions is that of the black swan.   Until Australia was discovered, swans had been found invariably to be white in a very large number of observations, and natural historians would have been justified in asserting, according to the principle suggested, the law that all swans are white ; and yet the law was false, for some swans in Australia are black. Instances of the second kind are plentiful in actual science. When a chemist makes a new compound, he often determines its melting-point or density ; as a result of a single measurement he will often be prepared to assert that its melting-point or density is higher (say) than that of water, and nobody will dream of doubting that the association he asserts is invariable or that subsequent measurements will lead to the same result.

These examples seem to prove that a large number of favourable instances, even if without exceptions, is neither sufficient nor necessary to establish a law. But at the same time they suggest what is the additional element required. We have omitted to take into consideration other laws closely similar to those that are under discussion. The chemist is certain that, in measuring the melting-point of a new compound, he is establishing an invariable relation, because from the examination of a great number of other compounds he has found that the density is an invariable property. On the other hand seventeenth-century naturalists ought to have regarded with suspicion a law that all swans are white (and probably they did actually so regard it) because the examination of other animals would have shown them that colour is by no means an invariable property, but is liable to vary very widely even among closely related species. In putting the matter as we did, the full evidence was not disclosed. The evidence for the invariable density of a new compound is not the single measurement of it, but the general law that all densities are invariable properties. This law is established by the observation, not of a single instance or of one or two, but of a very large number of instances, in none of which the relation has been found to fail. The evidence for the assertion of the law of the density of the new substance is really of exactly the same nature as that for the rising of the sun to-morrow.

This mode of expressing the matter is probably not quite correct ; for closer examination would show that it is difficult to regard the assertion that density (unlike colour) is an invariable property as a true law. It would be better to say that there are certain associations (such as that of density or melting-point with the other properties of a substance) which, if they occur at all, we expect to be invariable. In other words, we expect to find laws of certain forms, and if we find an observation which might be a particular instance of a law of one of these forms, we are much more ready to jump at once to the conclusion that this law is indeed true than we should be if the law, of which the observation would be a particular instance, is not of one of these forms. And one of the reasons why we expect such laws is that we have previously found a large number of them ; however, as we shall see presently, this is not the only reason.

THE ELASTICITY OF LAWS

But our answer is not complete yet. If this were all, I think we should feel far more uncertainty than we actually do feel about most of our laws. However many favourable instances we had observed, if we felt that a single unfavourable instance, if it occurred, would destroy the law, we should never be free from uneasiness. The contrary instance might occur ; we might go to our laboratory one morning and find that the density of some substance which we had measured the day before was now quite different. Our confidence in the law is largely based on the fact that such an unfortunate incident would not necessarily destroy our belief in the law.

This statement may be surprising. Surely if a law states that some relation is invariable ; and if, as we professed in Chapter III., we are going to be really strict in our interpretation of invariability, then a single contrary instance must destroy the law. For an association which has failed once, even if it has not failed a million times, is not strictly invariable. True ; but what exactly is the association we are asserting ? We are asserting that a certain density is invariably associated with a certain substance. If we find a new density we cannot maintain the invariable association if we attribute it to the same substance as that to which the old density was attributed. But why should we not attribute it to a new substance ? If we try the experiment over again and find that we do not get the same result as before, what is to prevent us avoiding any discrepancy between the two experiments by simply saying that they are not made on the same substance ?

Indeed this way out of the difficulty has been adopted implicitly in the case of the black swan. Since we have known of black swans, we do not say that there are not white swans ; we recognize two kinds of swans, one of which is black and the other white. Nor do we recognize any error in the assertion, by those who did not know of black swans, that all swans are white. All the swans that they knew anything about were white and have always remained white. The apparent difficulty arose only because the new birds were called swans. If we confine that term to the birds which were originally called swans, any law about swans is quite unaffected by the discovery of birds which resemble swans in some respects, but which, since they are not wholly the same, should not be called swans.

But, it may be urged, the case is not really parallel to that which we must suppose if we want to face the difficulty fairly. Black swans differ from white in other things than their colour, so that there is a reason quite apart from their unexpected colour for distinguishing them from white swans. Again, even after the discovery of black swans, white swans could still be found. But suppose—we will return now to the instance of density—that when we re-determined the density and found it changed, we could not detect any change in any other property of the substance and that we could not find a substance which, resembling this substance in all other respects, had the density found in the first experiment, could we then maintain the invariability of the association ? Well, it would doubtless be very awkward and men of science would get into a fever of excitement, but they *could* maintain their law. For the supposition that *nothing* had changed between the two experiments is impossible to realize ; the mere fact that a previous experiment had been made and that the second experiment had been made after the first is sufficient to make some change between the two. Of course our usual conception of a substance excludes the idea that such changes—a mere repetition of a measurement or the mere lapse of time—could change its properties and make it a new substance ; we should have to alter our conception of a substance. But that conception has been already altered so greatly since it was taken over from common sense that there would be no impossibility and no insuperable inconsistency in maintaining that, since we made the first experiment, the substance on which we made it had vanished from our ken and been replaced by some other substance, which might naturally enough have a different density.

Indeed we should have to maintain something of the kind, for, whatever we might do, the fact would remain that we have observed two densities which cannot be those of the same substance and cannot be asserted by the same law. Either we must include the two observations under different laws, or we must leave one (or both) of them outside laws altogether. We adopt this last alternative if we regard the first measurement simply as a mistake ; a mistake is something that is excluded necessarily from the subject-matter of science and to which, therefore, a law can have no reference. It is quite possible that, if such a case as we are imagining actually occurred, we should adopt this course; but it must be remembered that we might adopt the other and remove the discrepancy, not by rejecting the observation, but by stating two laws. Which alternative we shall adopt depends on all the circumstances, and here it is convenient to note why the observation of a very large number of favourable instances is important in the establishment of a law. If we have based a law on a large number of instances, and subsequently find other instances apparently discrepant, then, if, when we choose between the alternatives just mentioned, we reject the law, we place all these large number of observations outside the province of science. And this we are loath to do; we want to reduce as much as possible of our experience to order by means of laws, and the rejection of the whole of our past experience as one great "mistake," accords ill with that purpose. When we have ordered a very large number of instances by means of a law, we shall want to maintain that law at all hazards; and we shall be much more willing to introduce other laws to include instances apparently discrepant, and so to avoid the necessity of rejecting the material on which the original law was based, than we should be if we have only ordered a very small number of instances.

THE PURPOSE OF LAWS

It will be seen that in this discussion the question from which it started has almost been left out of account. We asked how we managed to establish laws, by the examination of our past experience, which were true also for future experience. The considerations that have been put forward suggest that this problem is not answered, but is hardly contemplated at all, in the actual discovery of laws. When we are seeking laws, we are only thinking about the experience that we have actually had ; and the problem which we seek to solve is one that has reference only to that experience. We seek to order the experience, to change it from a miscellaneous collection of apparently unconnected observations to a connected series of particular instances of a few wide principles. These principles by means of which, and in terms of which, we order our past experience are laws ; they state, as has been said so often before, associations between events and properties which have proved in our past experience to be invariable. It is because the associations have proved invariable throughout this experience that by means of them we can order the experience as many particular instances of a few principles. When our experience is increased by the addition of observations which *were* future but are now past, we seek once more to order in the same manner our increased volume of experience ; but in this increased volume all experience is of equal value, that which was future is in no way different from that which was past, for all is now past. It may happen that the order established for the original experience is equally valid for that which we now have ; the portion that is added can again be regarded as particular instances of the laws which were established as a result of the original experience. And if that happens, we have no reason to change our laws. But if that does not happen, if the laws established for the original experience do not prove valid when the volume of it is increased, then we have two alternatives. We may either reject altogether the added experience and say that it is not proper subject-matter for science, or we may alter slightly (or radically) our laws, so that they now order satisfactory both the old and the new. If we adopt the second alternative, the new laws propounded must still be such that they order the old experience, and they must therefore present some features of great similarity to the old laws. Which of the two alternatives we shall adopt depends upon which method leads to the most satisfactory ordering of the complete experience. For this reason the first alternative is never adopted if the second is available; for it means that we must leave unordered a portion of experience which we thought could be ordered.

This is, I believe, the attitude that is actually adopted by men of science in establishing laws. And if that is so, the conception of prediction does not enter into explicit consideration at all. We do not try to find laws that will predict; we only try to find laws that will order the experience that we have. It is possible to adopt that attitude because, although we know that we shall have future experience which has not been taken into consideration, that future experience can never force us to abandon the ground we have gained and to " disorder " the order that has been established. Whatever the experience may be, it will be possible either to order the increased volume of experience, or else to reject altogether from the subject-matter of science some portion of it, leaving only the remainder to be ordered.

THE VALUE OF LAWS

But to the practical man that attitude will not seem very satisfactory. It appears to deprive science of all objective value. If scientific laws are true, only because they can always be re-interpreted so that nothing can prove them false, then science is merely a childish game unworthy of the attention of any serious man. If, when science asserts that the sun will rise to-morrow, it only means that, if the sun does not rise, we propose to alter somewhat our laws of the solar system, science is mere trifling. What the plain man means by that assertion is that the sun *will* rise and that the expectation of its rising is a sound basis for the conduct of life ; he does not mean something that can be made true or false just as we please. It was all very well—I can hear an objector say—to insist at the beginning of the chapter that we can have no " knowledge " of future events ; it is undeniable that we have some kind of knowledge which we habitually use in our practical life ; and if the only kind of knowledge that science admits is a determination never to be proved wrong, then we must seek elsewhere for the information that undoubtedly is to be had.

Of course, I do not deny all this ; and now I shall try and show how the two points of view may be reconciled. Men of science, though they pay no direct attention to prediction, are not really indifferent to the success of their predictions, interpreted in accordance with the plainest common sense. If their predictions always failed, it would mean that each addition to experience would mean a new ordering of the whole. This ordering doubtless could be accomplished in some fashion, but it would have no value. The achievement of science would be like that of Penelope, who wove a cloth that she unravelled each night and started afresh each morning. If all our predictions were failures, we could, I suppose, continue our task of ordering experience, but no sane man would do so. Science is only worth while because it does make real progress. The ordering established for past experience is on the whole valid for future experience. The exceptions are comparatively few, and, even when they occur, it is found that the alteration of the order is so slight—it is often only a natural development of the old order—that the necessity for repeating the task is not wearisome. Time unravels, not the whole web, but only a few minor portions in which the shuttle has gone awry. Scientific laws do predict exactly in the manner which the plain man desires ; and it is really as necessary for the purposes of science that they should do so as for the purposes of practical life.

THE FUNDAMENTAL QUESTION

But why do they predict ? We return once again to the question which we cannot avoid. The final answer that I must give is that I do not know, that nobody knows, and that probably nobody ever will know. The position is simply this. We examine our past experience, and order it in a way that appears to us most simple and satisfactory ; we arrange it in a manner that is dictated by nothing but our desire that the world may be intelligible. And yet we find that, in general, we do not have to alter the arrangement when new experience has to be included. We arrange matters to our liking, and nature is so kind as to recognize our arrangement, and to conform to it ! If anyone asks, Why, what kind of answer can we possibly give ; how can we explain why the universe conforms to our intellectual desires ?

Here we inevitably touch upon profound problems, which lie far beyond the scope of this little book. I can only say that, for myself, none of the answers that have been offered seem satisfactory explanations, or even explanations at all ; they raise more questions and more difficult questions than they answer. But it may be well to draw attention to two considerations that have to be taken into account in any discussion of the matter. The first has been mentioned several times before. It must always be remembered that science does not attempt to order all our experience ; some part of it, and the part perhaps that is of most importance to us as active and moral human beings, is omitted altogether from the order. And it is very hard to say whether we omit it because we know that we cannot order it in the same manner as that which forms the

proper subject-matter of science, or because we feel
instinctively that, even if we could force it into such
an order, that order would not be appropriate to it.
I incline to the second alternative ; it seems to me
that there is something so fundamentally different
in the internal and external worlds (of Chapter II) that
we would not, even if we could, group them in the same
categories. But whichever alternative we adopt, it
remains equally difficult to explain why even the limited
part of experience which science takes as its province
conforms so closely to our desires, or why there should
be a part which can be selected so that it conforms.

The other consideration arises when it is asked who are
the " we," to whose intellectual desires nature conforms.
It is a grave difficulty, inherent in all the many attempts
to lay down rules whereby science may discover laws
valid for future experience, that they would indicate
that anybody who knew the rules could discover laws.
But that is not the fact ; it is not every one who has that
power. Indeed the fact seems to be precisely contrary.
Those who have professed the most intimate knowledge
of the rules, the great philosophers of science, such as
Bacon or Mill, have never been able to apply their rules
to the discovery of any law of the slightest value. Laws
have been discovered for the most part by people naïvely
innocent of all philosophical subtleties. The great man
of science, like the great poet or the great artist, is born
and not made ; like the artist he must train his faculties,
but training alone will not confer them. The vast majority
of mankind (a majority which includes a great many of
those who have done useful scientific work) cannot
discover laws, except in so far as they are helped, in a way
we shall notice immediately, by the previous work of an
infinitesimal minority. Either they cannot see order in
experience at all, or the order which they think they see
does not prove to be that to which nature is prepared to

conform ; they do not discover laws, or the laws that they discover predict falsely. It is only the great leaders of science who see the right order. They, and they only, can establish an order which satisfies their intellectual desires and yet find that it is valid for the future as well as for the past. They, and they only, are in such harmony with the universe that it obeys the dictates of their minds.

THE SIGNIFICANCE OF GENIUS

I fear this point of view will seem to some readers too mystical for their tastes. Nevertheless I would press it strongly on their attention. Of course I do not claim in the least that it explains why laws, devised even by the greatest of men, do predict, but it is necessary for the understanding of science, as much as for the understanding of art, to recognize that there are great men who surpass their fellows in some scarcely comprehensible manner. Science would not be what it is if there had not been a Galileo, a Newton or a Lavoisier, any more than music would be what it is if Bach, Beethoven and Wagner had never lived. The world as we know it is the product of its geniuses—and there may be evil as well as beneficent genius—and to deny that fact, is to stultify all history, whether it be that of the intellectual or the economic world.

But in one, as in the other, genius itself is too rare and too short-lived to achieve much by its unaided efforts. Great men—and this is particularly true of the greatest—achieve more by their influence than by their direct action. They change the world by enabling others to complete what they have themselves begun. And in no direction is this more true than in science. By far the greater part of scientific work has been done, and by far the greater number of laws discovered, by those of us who have not the remotest claims to genius or any but the very pedestrian talents of energy and application. But we are simply following in the footsteps of our masters. In Chapter III we noticed that there were standard forms of laws ; there are many laws, all quite distinct and ordering quite different groups of fact, which are yet obviously all of the same form. The laws asserting the properties of a substance provide a notable example ; there are many substances, but the laws which assert that there are such substances have all the same form. The properties of hydrogen, which are asserted by the law that there is such a thing as hydrogen, are quite different from the properties of iron, asserted by the law that there is such a thing as iron ; yet the laws are of the same form. Now once we have got the idea that there are laws of this form, it is a comparatively simple problem, which can be solved more or less according to fixed rules, to establish the laws of a new substance or, by finding new properties, to alter or augment an old one. And we do know now that laws of this particular form are among those to which nature will conform, and which can be usefully applied in prediction. The stroke of genius was that of the man who first suggested a law of that form ; once he had suggested it and showed that such a law is permanently valid, it was easy enough for others to take up the work and find others of the same form.

The discovery of the laws of substances is hidden in the darkness of the past ; they are among the ideas which we take over from common sense, and were invented by the unknown giants who laid the basis of human knowledge. But advances quite as important, the discovery of other forms of laws which have been used by the humbler folk who do the spade-work of science, such advances have occurred in historic times. Certain great men are recognized as the founders of certain branches of science, and if we inquire why they are so regarded, we shall usually find (but another reason will be found in the next chapter) that they were the first to establish a law of the form that is specially characteristic of that science. Thus of physics, numerical laws (which we shall discuss later) are especially characteristic ; Galileo was the first to establish a numerical law of the type of which almost all modern physics consists ; nine-tenths of the work of later physicists in the discovery of laws has been simply the extension of laws of Galileo's form to other fields of experience. Galileo may fairly be hailed as the founder of experimental physics. Other great men have so changed or amplified the form, that their work ranks as independent—Boyle, and Ampère may claim place in this class ; but again their fame rests largely on the discovery of a new type of law which has been simply applied elsewhere by lesser men. Of other sciences I am not competent to speak, but if Lavoisier is the founder of modern chemistry it is because he first established a law of the form that asserts chemical combination ; and if Linnæus is the founder of systematic botany, it is because he first established a law of the form that asserts the existence of a particular species of plant.

This then is really the solution of the main question of this chapter, as it faces the practising student of science. He believes that if he can discover a law of a certain form and order his experience in a certain way then that law will predict and nature will conform to that order. And so far, at least since the seventeenth century, his expectation has never been falsified ; I believe that in the history of modern science there is no instance of the abandonment of a type of law which has once been firmly established. Progress has been continuous ; it has consisted in the establishment of many laws of old types, and very occasionally, in the introduction of new types. Even when at first sight experience has contradicted expectation, it has always been possible (as in the example of the black swan) to remove the contradiction by resolving a law of one of these types into several laws of the same type, or by changing it to a law of another known type.

And what are these types ? To answer that question would be to expound all science ; I want only to encourage the reader to study science for himself and to find the types. But I have already indicated some of the more important types, such as the law of a substance, the law of a particular kind of animal and the numerical law ; and it has been urged that all these laws have the important common feature that the things between which they assert invariable associations are themselves interconnected by other laws. Those who have previously read in the philosophy of science will be surprised that the causal law is omitted, but the reasons of the omission were given in the previous chapter. In physics, at least, it is not an important type, though it possibly may be in other sciences, such as meteorology and medicine. And by omitting the causal laws, we can omit also all reference to the " Canons of Induction " which were supposed by an earlier generation to provide the one and only means for discovering scientific laws. They are futile, because the problem which they profess to solve is not one which has ever troubled any intelligent person. They tell us how, when we know that an event is the cause or effect of another, we may discover of which other event it is the cause or the effect ; as a matter of fact, the crudest common sense, applied in everyday life, serves for the purpose. We might similarly draw up canons for discovering of what substance a given property is a property, when it is once known that it is a property of some substance ; but here again the rules would be so obvious as not to be worth formulating. The problem of science is not to discover examples of laws when once we know what kind of law to look for ; it is to know for what kind of law to look. And that problem, as we have insisted before, is insoluble except by the genius which knows no rule.

CHAPTER V

THE EXPLANATION OF LAWS

THROUGHOUT the previous chapter I wrote as if the ordering of nature which is effected by laws was all that was necessary to satisfy our intellectual desires and so to fulfil the purpose of science. But really when we have discovered laws, we have fulfilled only part of the purpose of science. Even if we were sure that all possible laws had been found and that all the external world of nature had been completely ordered, there would still remain much to be done. We should want to *explain* the laws.

Explanation in general is the expression of an assertion in a more acceptable and satisfactory form. Thus if somebody speaks to us in a language we do not understand, either a foreign language or the technical language of some study or craft with which we are not familiar, we may ask him to *explain* his statement. And we shall receive the explanation for which we ask if he merely alters the form of his statement, so as to express it in terms with which we are familiar. The statement in its new form is more acceptable and more satisfactory, because now it evokes a definite response in our minds which we describe by saying that we understand the statement. Again we sometimes ask a man to *explain* his conduct ; when we make such a demand we are ignorant, or pretending to be ignorant, of the motives which inspired his action. We shall feel that he has offered a complete explanation if he can show that his motives are such as habitually inspire our own actions, or, in other words, that his motives are familiar to us.

But expressions, or the ideas contained in them, may be more acceptable and more satisfactory, on grounds other than their familiarity; and all explanations do not consist of a reduction of the less to the more familiar. Indeed it would seem that the explanations which, in the view of the man in the street, it is the business of science to offer do not involve familiar ideas at all. Thus we may expect our scientific acquaintances to *explain* to us *why* our water-pipes burst during a frost or *why* paint becomes dirty sooner in a room lit by gas than in one lit by electricity. We shall be told in reply that the bursting of the pipes is due to the expansion of water when it is converted into ice, and the blackening of the paint to the combination of the white pigment with sulphur present in coal-gas to form compounds that are dark, and not white, in colour. Now in these instances, the ideas involved in the explanation are probably less, and not more, familiar than those that they are used to explain. Many more people know that water-pipes burst during a frost than know that water (unlike most liquids) expands when it freezes; and many more know that their paint goes black than know that lead carbonate (one of the commonest white pigments) is converted by sulphur into black lead sulphide.

Why then do we regard our questions as answered ? Why do we feel that, when we have received them, the matter is better understood, and our ideas on it clearer and more satisfactory ? The reason is that the events and changes have been explained by being shown to be particular examples of a general law. Water *always* expands when it freezes, although it does not always burst household pipes; for it may not be contained in pipes or in any closed vessel. And lead carbonate *always* reacts with sulphur in the form present in coal-gas, even if it is not being used as a pigment. We feel that our experience is no longer peculiar and mysterious; it is only one instance of general and fundamental principles. It is one of the profoundest instincts of our intellectual nature to regard the more general principle as the more ultimately acceptable and satisfactory ; it is this instinct which led men first to the studies that have developed into science. In fact, what was called in the last chapter the " ordering " of experience by means of laws might equally well have been called the explanation of that experience. Laws explain our experience because they order it by referring particular instances to general principles ; the explanation will be the more satisfactory the more general the principle, and the greater the number of particular instances that can be referred to it. Thus, we shall feel that the bursting of the pipes is explained more satisfactorily when it is pointed out that the expansion of water when converted into ice explains also other common experience, for instance that a layer of ice forms first on the *top* of a pond and not on the bottom.

Doubtless there are other kinds of explanation ; but it is important for our purpose to notice that the explanations of common life often depend on these two principles —that ideas are more satisfactory when they are more familiar and also when they are more general ; and that either of these principles may be made the basis of an explanation.

When it is asked what is the nature of the scientific explanation of laws—and it is the purpose of this chapter to answer that question—it is usually replied that it is of the second kind, and that laws are explained by being shown to be particular examples of more general laws. On this view the explanation of laws is merely an extension of the process involved in the formulation of laws ; it is simply a progress from the less to the more general. At some stage, of course, the process must stop ; ultimately laws so general will be reached that, for the time being at least, they cannot be included under any more general laws. If it were found possible to include all scientific laws as particular instances of one extremely general and universal law, then, according to this opinion, the purpose of science would be completely achieved.

I dissent altogether from this opinion ; I think it leads to a neglect of the most important part of science and to a complete failure to understand its aims and development. I do not believe that laws can ever be explained by inclusion in more general laws ; and I hold that, even it were possible so to explain them, the explanation would not be that which science, developing the tendencies of common sense, demands.

The first point is rather abstruse and will be dismissed briefly. It certainly seems at first sight that some laws can be expressed as particular instances of more general laws. Thus the law (stating one of the properties of hydrogen) that hydrogen expands when heated seems to be a particular instance of the more general law, that all gases expand when heated. I think this appearance is merely due to a failure to state the laws quite fully and accurately, and that if we were forced to state with the utmost precision what we mean to assert by a law, we should find that one of the laws was not a particular case of the other. However, I do not wish to press this contention, for it will probably be agreed that, even if we have here a reference of a less general, to a more general law, we have no explanation. To say that all gases expand when heated is not to explain why hydrogen expands when heated ; it merely leads us to ask immediately why all gases expand. An explanation which leads immediately to another question of the same kind is no explanation at all.

WHAT IS A THEORY ?

How then does science explain laws ? It explains them by means of " theories," which are not laws, although closely related to laws. We will proceed at once to learn what a theory is, and how it explains laws.[1] For this purpose an example is necessary, even though its use involves entering more into the details of science than is our usual practice. A great many laws are known, concerning the physical properties of all gases ; air, coal-gas, hydrogen and other gases, differ in their chemical properties, but resemble each other in obeying these laws. Two of these laws state how the pressure, exerted by a given quantity of gas on its containing vessel, varies with the volume of the vessel, and with the temperature of the gas. Boyle's Law states that the pressure is inversely proportional to the volume, so that if the volume is halved the pressure is doubled ; Gay-Lussac's states that, at a constant volume, the pressure increases proportionally to the temperature (if a certain scale of temperature is adopted, slightly different from that in common use). Other laws state the relation between the pressure of the gas and its power of conducting heat and so on. All these laws are " explained " by a doctrine known as the Dynamical Theory of Gases, which was proposed early in the last century and is accepted universally to-day. According to this theory, a gas consists of an immense number of very small particles, called molecules, flying about in all directions, colliding with each other and with the wall of the containing vessel ; the speed of the flight of these molecules increases with the temperature ; their

[1] The reader should be warned that he must remember that the word " theory " is here used in a strictly technical sense of which the meaning is about to be explained. He must not attach to it any of the ideas associated with the word in ordinary language. In Chapter VIII some reference will be made to the use of " theory " in contra-distinction to " practice."

impacts on the walls of the vessel tends to force the walls outwards and represent the pressure on them ; and by their motion, heat is conveyed from one part of the gas to another in the manner called *conduction*.

When it is said that this theory explains the laws of gases, two things are meant. The first is that if we assume the theory to be true we can prove that the laws that are to be explained are true. The molecules are supposed to be similar to rigid particles, such as marbles or grains of sand ; we know from the general laws of dynamics (the science which studies how bodies move under forces) what will be the effect on the motions of the particles of their collisions with each other and with the walls ; and we know from the same laws how great will be the pressure exerted on the walls of the vessel by the impacts of a given number of particles of given mass moving with given speed. We can show that particles such as are imagined by the theory, moving with the speed attributed to them, would exert the pressure that the gas actually exerts, and that this pressure would vary with the volume of the vessel and with the temperature in the manner described in Boyle's and Gay-Lussac's Laws. In other words, from the theory we can deduce the laws.

This is certainly one thing which we mean when we say that the theory explains the laws ; if the laws could not be deduced from the theory, the theory would not explain the laws and the theory would not be true. But this cannot be all that we mean. For, if it were, clearly any other theory from which the laws could be deduced, would be equally an explanation and would be equally true. But there are an indefinite number of " theories " from which the laws could be deduced ; it is a mere logical exercise to find one set of propositions from which another set will follow ; and anyone could invent in a few hours twenty such theories. For instance, that the two propositions (1) that the pressure of a gas increases as the temperature increases (2) that it increases as the volume decreases, can be deduced from the single proposition that the pressure increases with increase of temperature and decrease of volume. But of course the single proposition does not explain the two others ; it merely states them in other words.   But that is just what logical deduction consists of ; to deduce a conclusion from premisses is simply to state the premisses in different words, though the words are sometimes so different as to give quite a different impression.[1] If all that we required of a theory was that laws could be deduced from it, there would be no difference between a theory which merely expressed the laws in different words without adding anything significant and a theory which, like the example we are considering, does undoubtedly add something significant.

It is clear then that when we say the theory explains the laws we mean something additional to this mere logical deduction ; the deduction is necessary to the truth of the theory, but it is not sufficient. What else do we require ?   I think the best answer we can give is that, in order that a theory may explain, we require it—to explain ! We require that it shall add to our ideas, and that the ideas which it adds shall be acceptable. The reader will probably feel that this is true of the explanation of the properties of gases offered by the dynamical theory.   Even if he did not know (and he probably does not know apart from what I have just told him) that the laws can be deduced from the theory, he would feel that the mere introduction of moving particles and the suggestion that the properties of a gas can be represented as due to their motion would afford some explanation of those properties.   They would afford some explanation,

[1] The reader should be warned that some people would dissent vehemently from this assertion.

even if the laws could not be deduced correctly ; they would then offer an explanation, although the explanation would not be true.

And this is, I believe, the reason why he would feel thus. Only those who have practised experimental physics, know anything by actual experience about the laws of gases ; they are not things which force themselves on our attention in common life, and even those who are most familiar with them never think of them out of working hours. On the other hand, the behaviour of moving solid bodies is familiar to every one ; every one knows roughly what will happen when such bodies collide with each other or with a solid wall, though they may not know the exact dynamical laws involved in such reactions. In all our common life we are continually encountering moving bodies, and noticing their reactions ; indeed, if the reader thinks about it, he will realize that whenever we do anything which affects the external world, or whenever we are passively affected by it, a moving body is somehow involved in the transaction. Movement is just the most familiar thing in the world ; it is through motion that everything and anything happens. And so by tracing a relation between the unfamiliar changes which gases undergo when their temperature or volume is altered, and the extremely familiar changes which accompany the motions and mutual reactions of solid bodies, we are rendering the former more intelligible ; we are explaining them.

That is to say, the explanation of laws offered by theories (for this example has been offered as typical) is characteristically explanation of the first of the two kinds with which the chapter started. It is explanation by greater familiarity, essentially similar to that offered when a statement is translated from an unknown to a known language. This conclusion may be surprising, and indeed it is not that generally advanced. Before developing our view further, it will be well to examine the matter from another point of view.

DIFFERENCE BETWEEN THEORIES AND LAWS

It was stated before that it has been usually held that the explanation of laws consists in showing that they are particular examples of more general laws. If this view were applied to the example under discussion, it might be urged that the dynamical theory explains the properties of gases because it shows that they are particular examples of the laws of dynamics ; the properties of gases are explained because they are shown to be the consequences of the subjection of the molecules, of which the gases consist, to the general laws of all moving bodies. Here, it might be said, is the clearest possible instance of explanation by generalization, a simple extension of the process involved in the discovery of laws.

But, against this view, it must be pointed out that the most important feature of the theory is not that it states that molecules are subject to dynamical laws, but that which states that there are such things as molecules, and that gases are made up of them. It is that feature of the theory which makes it a real explanation. Now this part of the theory is not a particular instance of any more general law ; indeed it is not a law or anything that could be an instance of a law. For it is not, according to the criterion laid down in Chapter II, part of the proper subject-matter on which science builds its foundations. Molecules are not things which we can see or feel ; they are not, like the ordinary material bodies to which the laws of dynamics are known to apply, objects discernible to direct perception. We only know that they exist by inference ; what we actually observe are gases, varying in temperature and pressure ; and it is only by these variations that we are led to suspect the existence of the molecules. We may apply once more our fundamental test of universal agreement which serves to distinguish the objects concerned in laws from any others. If somebody denied the existence of molecules, how could we prove him wrong ? We cannot show him the molecules ; we can only show him the gases and expound the theory ; if he denied that the theory proved the existence of the molecules, we should be powerless. We cannot prove by his actions that he is perverse or deluded ; for his actions will be affected only by the properties of gases, which are actually observed, and not by the theory introduced to explain them. Actually the dynamical theory of gases has been denied by men of science of high distinction. Usually the denial was based partly on the assertion that the laws of gases could be deduced accurately from the theory, but it has often been accompanied by the contention that, even if they could be deduced accurately, the theory was not true, and not worthy of acceptance. No denial of that case would be possible if the theory were indeed a law.

We conclude therefore—and the conclusion is vital to the view of science presented here—that a theory is not a law, and consequently, that the explanation afforded by a theory cannot simply be the explanation by generalization which consists in the exhibition of one law as a particular instance of another. It does not follow that theories have nothing to do with laws, and that it is immaterial for the theory that the laws of dynamics are true, and of very great generality. We shall see presently that this feature is of great importance. But it does not involve that the theory is itself a law.

THE VALUE OF THEORIES

After this protest against a dangerous misunderstanding, let us return and develop further our view of theories,

So far the truth of a theory has been based on two grounds ; first, that the laws to be explained can be deduced from it ; second, that it really explains in the sense that has been indicated. But actually there is in addition, a third test of the truth of a theory, which is of great importance ; a true theory will not only explain adequately the laws that it was introduced to explain ; it will also predict and explain in advance laws which were unknown before. All the chief theories in science (or at least in physics) have satisfied this test ; they have all led directly to the discovery of new laws which were unsuspected before the theory was proposed.

It is easy to see how a theory may predict new laws. The theory, if it is worthy of consideration at all, will be such that the old laws can be deduced from it. It may easily be found on examination that not only these laws, but others also can be deduced from it ; so far as the theory is concerned, these others differ in no way from the known laws, and if the theory is to be true, these laws that are consequences of it must be true. As a matter of fact, it is very seldom that a theory, *exactly* in its original form, predicts any laws except those that it was proposed to explain ; but a very small and extremely natural development of it may make it predict new laws. Thus, to take our example, in order to explain the laws (Boyle's and Gay-Lussac's) to which the theory was originally applied, it is unnecessary to make any assumption about the size of the molecules ; those laws can be deduced from the theory whatever that size (so long as it is below a certain limit) and the assumption was at first made for simplicity that the molecules were mathematical points without any size at all. But obviously it was more natural to assume that the molecules, though extremely small,[1] have some size and once that assumption is made,

---

[1] If a drop of water were magnified to the size of the earth, the molecules would be about the size of cricket balls.

laws are predicted which had not been discovered at
the time and would never have been suspected apart
from the theory. Thus, it is easy to see that, if the
molecules have a definite size, the behaviour of a gas,
when the number of molecules contained in a given vessel
is so great that the space actually occupied by the molecules is nearly the whole of the space in the vessel, will
be very different from its behaviour when there are
so few molecules that practically all that space is unoccu-
pied. This expectation, a direct result of the theory, is
definitely confirmed by experiments which show a change
in the laws of a gas when it is highly compressed, and all
its molecules forced into a small volume.

This test of predicting new and true laws is always
applied to any theory when it is proposed. The first
thing we do when anyone proposes a theory to explain
laws, is to try to deduce from the theory, or from some
slight but very natural development of it, new laws,
which were not taken into consideration in the formulation
of the theory. If we can find such laws and prove by
experiment that they are true, then we feel much more
confidence in the theory ; if they are not true, we know
that the theory is not true ; but we may still believe that
a relatively slight modification will restore its value. It
is in this way that most new laws are actually suggested
for the purposes discussed in the previous chapter. At
the present time, in the more highly developed sciences,
it is very unusual for a new law to be discovered or suggested simply by making experiments and observations
and examining the results (although cases of this character
occur from time to time) ; almost all advances in the
formulation of new laws follow on the invention of theories
to explain the old laws. Indeed it has been urged that
the only use of theories is thus to suggest laws among
which some will be found to be true. This opinion has
been much favoured by philosophers and mathematicians

and has always been accompanied by the opinion that it is the end and object of science to discover laws It has also been professed (especially at the end of the nineteenth century) by people who know something about science and actually practised it ; but I think that these people only professed the view because they were afraid what the philosophers might say if they denied it. At any rate, for myself, I cannot understand how anybody can find any interest in science, who thinks that its task is completed with the discovery of laws.

For the explanation of laws, though it is formally quite a different process from the discovery of laws, is in its object merely an extension of that process. We seek to discover laws in order to make nature intelligible to us ; we seek to explain them for exactly the same reason. The end at which we are aiming in one process as in the other is the reconciliation with out intellectual desires of the perceptions forced on us by the external world of nature What possible reason can be given for attaching immense importance to one stage in the process and denying all intrinsic importance to another ? Surely so long as anything remains to be explained it is the business of science to continue to seek explanations.

THE INVENTION OF THEORIES

And here again arises obviously a question very similar to that discussed in the previous chapter. A theory, it has been maintained, is some proposition which satisfies these conditions : (1) It must be such that the laws which it is devised to explain can be deduced from it ; (2) it must explain those laws in the sense of introducing ideas which are more familiar or, in some other way, more acceptable than those of the laws ; (3) it must predict new laws and these laws must turn out to be true. Of course we have to ask now how such theories are to be found ; we might find theories to satisfy the first two conditions by the exercise of sufficient patience in a process of trial and error ; but how can we possibly be sure that they will satisfy the third ? The answer that must be given will be clear to the reader if he has accepted the results of the previous discussion. There cannot be, from the very nature of the case, any kind of rule whereby the third condition may certainly be satisfied ; the meaning of the condition precludes any rule. Actually theories are always suggested in view of the first two conditions only ; and actually it turns out that they often fulfil the third. And once again they most often turn out to fulfil the third when they are suggested by those exceptional individuals who are the great men of science. It is when the theory seems to *them* to explain the laws, when the ideas introduced by it appear to *them* acceptable and satisfactory, that nature conforms to their desires, and permits to be established by experiment the laws which are the direct consequence of those ideas.

The form in which that statement is put may appear rather extravagant, and we shall return later and consider some questions which it seems to raise. But the general view that true theories are the expression of individual genius will probably seem less paradoxical than the view put forward in the previous chapter, that true laws also contain a personal element. The difference is indicated by the words that are used ; we speak of the " discovery " of a law, but of the " invention " of a theory. A law, it is implied, is something already existing which merely lies hidden until the discoverer discloses it ; a theory, on the other hand, does not exist apart from the inventor ; it is brought into being by an effort of imaginative thought. I do not think that distinction will bear examination ; it seems to me very.difficult to regard laws and not theories as something existing independently of investigation and wholly imposed by the external world, or theories and not laws as the product of the internal world of the intellect. For both theories and laws derive their ultimate value from their concordance with nature and both arise from mental processes of the same kind. Moreover, as has been suggested already, in the more highly developed sciences of to-day theories play a very large part in determining laws ; they not only suggest laws which are subsequently confirmed by experimental investigation, but they also decide whether suggested laws are or are not to be accepted. For, as our discussion in the previous chapter showed, experiment alone cannot decide with perfect definiteness whether or no a law is to be accepted ; there are always loopholes left which enable us to reject a law, however much experimental evidence may suggest it and enable us to maintain a law (slightly modified) even when experimental evidence seems directly to contradict it. An examination of any actual science will show that the acceptance of a law is very largely determined by the possibility of explaining it by means of a theory ; if it can be so explained, we are much more ready to accept it and much more anxious to maintain it than we should be if it were not the consequence of some theory. Indeed many laws in science are termed " empirical " and regarded with a certain amount of suspicion ; if we inquire we find that an empirical law is simply one of which no theoretical explanation is known. In the science of physics at least, it would almost be more accurate to say that we believe our laws because they are consequences of our theories than to say that we believe our theories because they predict and explain true laws !

On such grounds I reject the view (though it is generally prevalent) that laws are any less the product of imaginative thought than are theories. The problem why nature conforms with our intellectual desires arises just as clearly with one as with the other. Nevertheless it is doubtless true that the personal and imaginative element is more obvious and more prominent in theories than in laws. One aspect of this difference has been noted already ; the acceptance of a theory as true does involve a personal choice in a way that a law does not. Different people do differ about theories ; they can choose whether or no they will believe them ; but people do not differ about laws ;[1] there is no personal choice ; universal agreement can be forced. Again, if we look at the history of science, we shall find that the great advances in theory are more closely connected with the names of the great men than are the advances in laws. Every important theory is associated with some man whose scientific work was notable apart from that theory ; either he invented other important theories or in some other way he did scientific work greatly above the average. On the other hand there are a good many well-known laws which are associated with the names of men who, apart from those particular laws, are practically unknown ; they discovered one important law, but they have no claim to rank among the geniuses of science.[2] That fact seems to indicate that a greater degree of genius is needed to invent true theories than to discover true laws.

The same feature appears in the early and prehistoric stages of science. Science, as we have seen, originally took over from common sense laws which had been already elaborated ; and although it has greatly refined and elaborated those laws and has added many new types, it has never wholly abandoned the laws of common sense. Modern science depends as much as the crudest common sense on the notion of a " substance " (a notion

[1] Except in so far as people may refuse to admit a law or to regard it as anything but empirical, because it is not in accordance with some theory.   But then they admit that the laws describe the facts rightly ; they only suggest that some other and equally accurate way of describing them would be preferable.

[2] In case this book falls into the hands of some expert physicist, I would suggest that examples may be found in the laws of Stefan, Dulong and Petit, or Bode.

which, as we have seen, implies a law), on the notion of the succession of events in time and the separation of bodies in space and so on. But science has abandoned almost all the theories of pre-scientific days. For there were and are such non-scientific theories ; and it is because the plain man has theories of his own, just as much as the most advanced man of science, that it has not been necessary to occupy a larger space in explaining exactly what a theory is ; the reader will probably have recognized at once something familiar in the kind of explanation which the dynamical theory of gases offers. The most typical theories of the pre-scientific era were those which explained the processes occurring in nature by the agencies of beings analogous to men—gods, fairies, or demons. The " Natural Theology " of the eighteenth century which tried to explain nature in terms of the characteristics of a God, known through His works, was a theory of that type ; in the features which have been described as essential to theories it differed in no way from that which we have discussed. But all such theories have been abandoned by science ; the theories that it employs are of a type quite unknown before the seventeenth century.[1] In respect of theories science has diverged completely from common sense ; and the divergence can be traced accurately to the work of a few great men. Common sense is therefore more ready to accept theories rather than laws as the work of individual genius.

But while I accept fully the view that the formulation of a new theory, and especially of a new type of theory,

[1] An exception is often made in favour of Lucretius, who wrote about 70 B.C. But my own opinion is that moderns, with their fuller knowledge, read into his works ideas which never entered the head of their author. I do not think that (as Mr. Wells maintains in his " Outline of History ") it was merely the barrenness of the soil on which his seed fell that prevented it blossoming into fruit. The sterility of his ideas, contrasted with those of Galileo and Newton, was inherent in them.

is a greater achievement than the formulation of a new law, I cannot admit that the two processes are essentially different. As Galileo was the founder of experimental physics, so Newton was the founder of theoretical physics ; as Galileo first introduced the type of law which has become most characteristic of the science, so Newton introduced first the characteristic type of theory. And of the two Newton is rightly judged by popular opinion to have been the greater man. But it is not rightly recognized how great was the achievement of Galileo ; indeed his fame is usually associated with things—the observation of the isochronism of the pendulum, or his fight with clericalism over the Copernican theory— other than with his greatest service to science. It is his establishment of the first experimental numerical law that constitutes his highest claim to greatness, and that law was as much an expression of his personality as the theory of Newton.

THE ANALOGIES OF THEORIES

Mention has just been made of " types " of theories. There are such types, just as there are types of laws, and they play the same rôle in permitting lesser men to complete and extend the work of the greater. Once a theory of a new type has been invented and has been shown to be true in the explanation of laws, it is naturally suggested that similar theories may prove equally successful in the explanation of other laws. And on the whole the suggestion has proved true. In each branch of science there are certain very broad and general theories which have been invented by the founders of that branch ; subsequent development of that branch usually consists largely in the ampliation and slight modification of such fundamental theories by investigators, many of whom could never have themselves laid the foundation. Indeed, the investigator often feels that in finding an explanation for the laws that he has discovered he has little more latitude than in discovering those laws ; it is perfectly clear from the outside what kind of theory he must seek, just as it is clear what kind of law he must seek.

Thus, it may be stated broadly[1] that from 1700 to 1870 all physical theories were of a single type of which the dynamical theory of gases which we have used as an example provides an excellent instance. They were all " mechanical " theories. In our example, the behaviour of gases is explained by an analogy with a piece of mechanism, a set of moving parts reacting on each other with forces which determine and are determined by the motion. That feature is common to all the mechanical theories which played so great a part in the older physics and are still prominent in the newer ; they explain laws by tracing an analogy between the system of which the laws are to be explained and some piece of mechanism. Once it was realized that such theories were likely to turn out to be true, the task of inventing theories was greatly simplified ; it often became simply that of devising a piece of mechanism which would simulate the behaviour of the system of which the laws were to be explained.

But all scientific theories are not mechanical. In physics it is the admission of theories that do not fall within this class which distinguishes the newer from the older study. And in other branches of science (except where they are obviously founded on physics) theories of other kinds are the rule. For instance, the theory of evolution, proposed to explain the diversity and yet the resemblances of different kinds of living beings is

---

[1] A very important exception must be made of the purely mathematical theories, such as those of Fourier and Ampère. Some indication of the nature of these theories is given in Chapter VII.

not mechanical ; it does not trace an analogy between the production of such beings and the operation of a piece of mechanism. Can we find any feature that is common to all theories that have proved to be true, or must we (as in the case of laws) rest content with several distinct but well-defined types which have all proved successful and yet display no common characteristic ?

I think we can find such a feature. The explanation offered by a theory (that is to say, the part of the theory which does not depend simply on the deduction from it of the laws to be explained) is always based on an analogy, and the system with which an analogy is traced is always one of which the laws are known ; it is always one of those systems which form part of that external world of which the subject-matter of science consists. The theory always explains laws by showing that if we imagine that the system to which those laws apply consists in some way of other systems to which some other known laws apply, then the laws can be deduced from the theory. Thus our theory of gases explains the laws of gases on the analogy of a system subject to dynamical laws. The theory of evolution explains the laws involved in the assertion that there are such-and-such living beings by supposing that these living beings are the descendants of others whose characters have been modified by reaction to their surroundings in a manner which is described by laws applicable to living beings at the present day. Again the immense theory involved in the whole science of geology explains the structure of the earth as it exists to-day by supposing that this structure is the result of the age-long operation of influences, the action of which is described by laws observable in modern conditions. In each case the " explaining " system is supposed to operate according to known laws, but it is not a system of which those laws can be asserted as laws, because it is, by the very supposition underlying the theory, one which could never be observed—either because it is too small or too remote in the past or for some similar reason—and therefore does not form part of the proper subject-matter of science.

It is because the explanation offered by a theory is always based on an analogy with laws that the distinction between laws and theories has been so often overlooked. The statement of the dynamical theory of gases about the properties and behaviour of molecules, or of the theory of evolution about the process whereby the existing species of living beings came into existence, is so similar to the statement, asserted by a law, about the properties of actual mechanical systems or about the changes that are proceeding in existing species that the vital difference between the two is forgotten. The statement asserted by a law can be proved by direct perception ; it states something which can be observed and which can be the subject of universal assent. The statement involved in the theory cannot be proved by direct perception, for it does not state anything that can be or has been observed. The failure to observe this distinction and the consequent failure to give to theories their true place in the scheme of science is the cause of most of the misunderstandings that are so widely prevalent concerning the nature and objects of science. For it has been admitted that, though the discovery of laws depends ultimately not on fixed rules but on the imagination of highly gifted individuals, this imaginative and personal element is much more prominent in the development of theories ; the neglect of theories leads directly to the neglect of the imaginative and personal element in science. It leads to an utterly false contrast between " materialistic " science and the " humanistic " studies of literature, history and art.

SCIENCE AND IMAGINATION

At the risk of wearying the reader by endless repetition I have insisted on the fallacy of neglecting the imaginative element which inspires science just as much as art. If this book is to fulfil any useful purpose that insistence is necessary. For it is my object to attract students to science and to help them to understand it by showing them from the outset what they may expect from it. It is certain that one of the chief reasons why science has not been a popular subject in adult education, and is scarcely recognized even yet as a necessary element of any complete education, is the impression that science is in some way less human than other studies. And for that impression men of science are themselves more to blame than anyone else ; in a mistaken endeavour to exalt the certainty of their knowledge they deliberately conceal that, like all possible knowledge, it is personal. They exhibit to the outside world only the dry bones of science from which the spirit has departed.

It is true that it is less easy for the beginner to grasp the imaginative element in science than in some other studies. A larger basis of mere information is perhaps required before it becomes apparent. And, of course, he can never hope to share himself the joy of discovery ; but in that respect he is no worse off than many who make of science their life-work. But he can, if he will take the trouble, appreciate the discoveries of others and experience at second-hand the thrill of artistic creation. For those who have the necessary knowledge, it is as exciting to trace the development of a great scientific theory, which we could never have developed ourselves, as it is to read great poetry or to hear great music which we ourselves could never have written. But I must admit that the books on science are few which make it easy for the beginner to share that experience. And so, although I can hardly hope that I shall succeed where so many writers have failed, and although the attempt transgresses the strict limit of an introduction, I should like to try to tell again the familiar story of one of the most wonderful romances of science—the story of Newton and the apple ![1]

The early chapters of the story must be greatly abbreviated. Copernicus and Kepler, a century before Newton, had shown clearly what were the paths in which the planets move about the sun and the satellites, such as our moon, about the planets. It does not seem clear whether anyone before Newton had thought of inquiring *why* they should move in such paths, or had ever contemplated the possibility of explaining the laws which Kepler had laid down. In science, as in many other things, it is often much harder to ask questions than to answer them. People might have said, and many probably did say : The planets have to move somehow ; the paths Kepler describes are quite simple ; why should not the planets move in them ? It is as ridiculous to ask why they so move as to ask why a man's hair is yellow or brown or red, and not blue or green. The mere conception of explaining the paths of the planets was itself an immense achievement.

And we can see now what suggested it to Newton. Some sixty years before Galileo had, for the first time, discovered some of the laws which govern the motions of bodies under forces. He had shown that, in some simple instances at least, there were such things as " laws of dynamics." The idea occurred to Newton, May not the movements of the planets and their satellites be subject to just such laws of dynamics as Galileo had discovered for the ordinary bodies which we see and handle. If so,

[1] Of course the apple may be mythical—like all the historical objects of our childhood. And it is impossible to be certain what Newton really thought. But his thought might have followed the line suggested here.

we ought to be able to find a set of forces such that if they act on the planets according to Galileo's laws, the planets will move as Kepler has shown that they do move. That seems to us very obvious now ; but it was not obvious then. Galileo, as far as we know, never thought of it ; nor did anyone in the two generations between him and Newton. And perhaps one reason why nobody thought of it was that they realized instinctively that if they had thought of it they could have got no further. To-day any clever schoolboy could solve the next problem which presents itself, namely that of finding *what* forces, acting according to Galileo's laws, would make the planets move as they do ; but that is only because Newton has shown him the way. In order to solve this problem which seems to us now so easy, Newton had to invent modern mathematics ; he had to make a greater advance in mathematics than had been made in all the time since the high-water of Egyptian civilization. This achievement of his was quite as wonderful as any other ; but as it was not characteristically scientific (in the modern sense) it may be left on one side here.

So he solved his problem. He showed that the planets can be regarded as subject to Galileo's laws, and that the force on a planet must be directed towards the sun, and that on the moon towards the earth ; and that these forces must vary in a certain simple way with the distance between planet and sun or moon and earth. The moon follows the course she does because there is a pull between it and the earth just as, when a stone is whirled round at the end of a string, there is a pull between the stone and the hand.

And now—as I like to think—he had ended his labour. He realized that he had made a stupendous discovery, which must revolutionize, as it actually has done, the whole science of astronomy. He had shown that the laws of dynamics apply to planets as well as to ordinary bodies on the earth, that the planets were subject to forces, and had determined what those forces were. What more could he do ?   What explanation of his result could be offered or even demanded.   He had ordered the solar system, and who could be so foolish as to ask why the order was that which he had found and no other ? But after his morning's work which finally completed his forthcoming treatise on the matter, he sat in his orchard and was visited by some of his friends from Cambridge. Perhaps they too were natural philosophers and he talked to them about what he had been doing ; but it is more likely that they sat idle, talking the thin-blooded intellectual scandal that must have always flourished in academic society, while Newton played with the historic kitten.

And then the apple fell from the tree.   Newton was suddenly silent in a reverie ; the kitten played unheeded with the fallen fruit ; and his friends, used to such sudden moods, laughed and chattered.   After a few moments he must have paper ; he rushes to his desk in the house ; scribbles a few hasty figures ; and now the theory of gravitation is part of the structure of the universe.   The falling apple, a trivial incident which he had seen a thousand times before, had loosed a spring in his mind, set unconsciously by all his previous thought.   He had never consciously asked himself why the moon was pulled towards the earth until, in an instant of revelation, the apple appeared to him not " falling " (the meaningless word he had always used before), but *pulled towards the earth*.   The idea flashed on him quicker than it could be spoken.   If both the moon and the apple are pulled towards the earth, may they not be pulled by the same force ?   May not the force that makes the apple " fall " be that which restrains the moon in its orbit ?   A simple calculation will test the idea.   He knows how far the moon is from the earth and how the force on it varies with the distance and with the size of the moon.   If the moon were brought to the surface of the earth and reduced to the size of the apple would the force on it be such that it would fall with the speed of the apple ?   The answer is, Yes !¹   The motion of the planets is therefore explained both by generalization and by familiarity. That motion is merely one particular instance of a general principle of which the very familiar fall of heavy objects is another.

What I want to impress on the reader is how purely personal was Newton's idea.   His theory of universal gravitation, suggested to him by the trivial fall of an apple, was a product of his individual mind, just as much as the Fifth Symphony (said to have been suggested by another trivial incident, the knocking at a door) was a product of Beethoven's.   The analogy seems to me exact. Beethoven's music did not exist before Beethoven wrote it, and Newton's theory did not exist before he thought of it.   Neither resulted from a mere discovery of some-thing that was already there ; both were brought into being by the imaginative creation of a great artist.   How-ever there is one apparent difference ; Beethoven having

¹ The reader who knows the story—and who does not ?—will see that here I deviate widely from history.   Newton did not know how far the moon was from the earth ; current estimates were wrong ; and at first he was therefore doubtful of his theory.   But when the distance was measured more accurately, he found it agreed perfectly with his theory.   I hesitate to suggest that it was Newton's theory that had changed the distance !

I feel, too, that some people will think that I must be very antiquated in my knowledge if I glorify Newton so greatly when the daily Press has been assuring us lately that his ideas have been completely over-thrown by Einstein.   This is not the place to discuss what Einstein has proved.   I admire his work as much as anyone, but he has not invalidated in the smallest degree the great discovery of Newton which is discussed in the text.   It is still as certain as ever it was that the fall of the apple and the " fall " of the moon are merely two examples of the very same fundamental principle ; and it is as certain as ever that the motions of the planets are subject to the same laws as those of terrestrial bodies.   What is now not quite certain is whether Galileo's laws are strictly applicable to circumstances very different to those of the experiments by which he proved them.

conceived his symphony had no need to test it in order to discover if it was " right," while Newton had to compare the results of his theory with the external world, before he was sure of its true value.   Does not this show that Newton's achievement was not so perfectly personal and imaginative as Beethoven's ?

I do not think so.   First, Beethoven's work *had* to be tested ;   the test of artistic greatness is appeal to succeeding generations free from the circumstances in which the work was conceived ;   it is very nearly the test of universal agreement.   But it is another point of view I want to emphasize here.   It is said that Newton's theory was not known to be true before it was tested ;   but Newton knew that it was true—of that I am certain.   To our lesser minds there seems no imperative reason why the force on the moon and the force on the apple should be related as closely as the theory of gravitation demands, merely because it would be so delightfully simple if they were ;   but Newton probably felt no doubt whatever on the matter.   As soon as it had occurred to him that the fall of the apple and the fall of the moon might be the same thing, he was utterly sure that they were the same thing ;   so beautiful an idea *must* be true.   To him the confirmation of numerical agreement added nothing to the certainty ;   he examined whether the facts agreed with the object of convincing others, not himself.   And when the facts as he knew them did not agree, we may be sure that his faith in the theory was in no way shaken ;   he *knew* that the facts must be wrong, but he had to wait many years before evidence of their falsity was found which would appeal to those who had not his genius and could not be sure of harmony between their desires and the external world.

ARE THEORIES REAL ?

And here we come to our last question. I have been at pains to distinguish theories from laws, and to insist that theories are not laws. But if that contention is true, are not theories deprived of much of their value ? Laws, it may be said, are statements about real things, about real substances (such as iron), about real objects (such as the earth or the planets or existing living beings). Laws are valuable because they tell us the properties of these real objects. But if theories are not laws, and if the statements they make are about things that cannot ever be the subject of laws, do they tell us about anything real ? Are the molecules (by means of which we explain the properties of gases) or the countless generations of unknown animals and plants (by means of which we explain the connexions between known animals and plants) or the forces on the planets (by means of which we explain their orbit)—are these molecules and animals and forces mere products of our fantasy, or are they just as real as the gases and the animals the laws of which they are led to explain ? Are theories merely explanatory, are they like the fairy tales by means of which our ancestors explained to themselves the world about them, are they like the tales we often tell to our children with the same object of explanation, or are they truly solid fact about the real things of the world ?

That may seem a simple question to which a plain answer, Yes or No, might be given ; but in truth it raises the most profound and abstruse problems of philosophy and really lies without the scope of this book. Our object is to discover what science is ; we have learnt what laws and theories are, and what part they play in science ; it is not directly part of our purpose to discuss what value all this elaboration has when it is achieved. But in a book of this kind it would be wrong to leave the question with no answer ; and I will therefore explain how the matter appears to me, although I know that many other people would give different answers.

I should reply to the questioner by asking him what he means by " real " and why he is so sure that a piece of iron, or a dog, is a real object. And the answer that I should suggest to him is that he calls these things real because they are necessary to make the world intelligbile to him ; and that it is because they are necessary to make the world intelligible to him that he resents so strongly (as he will if he is a plain man) the suggestions that some philosophers have made that these things are not real. It is true that these suggestions are often not interpreted rightly, and that what the philosophers propose is not so absurd as appears at first sight ; but the fact remains that these ideas are of supreme importance to him in making the world intelligible, and that he dislikes the notion that they are in any sense less valuable than other ideas which, for him at least, do not make the world so intelligible. The invariable associations which are implied by the use of the ideas " iron " and " dog " are extremely important in all his practical life ; it is extremely important for him that a certain hardness and strength and density and so on are invariably associated in the manner which we assert when we say that there is iron, and that a certain form and sound and behaviour are invariably associated in the way that we assert when we say that there are dogs. When the plain man says iron and dogs are real objects he means (I suggest) to assert that there are such invariable associations, that they are extremely important, and that they are rendered intelligible only by the assertion that there is iron and there are dogs.

If we accept this view it is clear that we must answer in the affirmative the question from which we started. Theories are also designed to make the world intelligible to us, and they play quite as important a part as do laws in rendering it intelligible. And if anything is real that renders the world intelligible, then surely the ideas of theories—molecules and extinct animals and all the rest of it—have just as much claim to reality as the ideas of laws.

But my questioner will almost certainly not be satisfied with that answer ; it will seem to him to shirk just the question that he wants to raise. He will feel that the view that reality is merely what leads to intelligibility deprives reality of all its importance ; if science is merely an attempt to render the world intelligible, in what does it differ from a fairy tale—which has often the same object ? Or—to put the matter in a different way— intelligibility is a quality that depends on the person who understands ; one person may find intelligible what another may not. Reality on the other hand is, by its very meaning, something independent of the person who thinks about it. When we say that a thing is real we do not mean that it is peculiarly suited to our understanding ; we mean much more that it is something utterly independent of all understanding ; something that would be the same if nobody ever thought about it at all or ever wanted to understand it.

I think the essence of this objection lies in the sentence : " One person may find intelligible what another may not." When we feel that science is deprived of all value by being likened to a fairy tale, our reason is that different people like different fairy tales, and that one fairy tale is as good as another. But what if there were only one possible fairy tale, only one which would explain the world, and if that one were intelligible and satisfactory to every one ? For that is the position of science. There have been many fairy tales to explain the world ; every myth and every religion is (in part at least) a fairy tale with that object. But the fairy tale which we call science differs from these in one all-important feature ; it is the fairy tale which appeals to every one, and the fairy tale which nature has agreed to accept. It is not you and I and the man round the corner who find that the conception of iron makes the world intelligible, while the people in the next street do not ; in this matter every single living being in the world (so far as we can ascertain his opinion) agrees with us ; they all accept our fairy tale and agree that it makes the world intelligible. And nature accepts it too ; the law that there is iron enables us to predict, and nature always agrees with our predictions. There is no other fairy tale like this ; there is none that denies that there is iron, a substance with invariably associated properties, which is acceptable to every one and which predicts truly. It is just because our fairy tale is capable of the universal agreement which we discussed in Chapter II that we distinguish it from all other fairy tales and call it solid fact. Nevertheless the fact remains that its value for us is that of other fairy tales, namely that it makes the world intelligible.

And now let us turn again to theories. Here, it is true, we cannot apply directly the criterion of universal assent. There is actually much more difference of opinion concerning the value of theories than there is concerning the value of laws ; and it is impossible to force an agreement as it can be forced in the case of laws. And while that difference of opinion persists we must freely admit that the theory has no more claim on our attention than any other ; it is a fairy tale which may be true, but which is not known to be true. But in process of time the difference of opinion is always resolved ; it vanishes ultimately because one of the alternative theories is found to predict true laws and the others are not. It is for this reason that prediction by theories is so fundamentally important ; it enables us to distinguish between theories and to separate from among our fairy tales that one which nature is prepared to accept and can therefore be transferred from the realm of fantasy to that of solid fact.   And when a theory has been so transferred, when it has gained universal acceptance because, alone of all possible alternatives, it will predict true laws, then, although it has purpose and value for us because it renders the world intelligible, it is so clearly distinguished from all other attempts to achieve the same purpose and to attain the same value that the ideas involved in it, like the ideas involved in laws, have the certainty and the universality that is characteristic of real objects. A molecule is as real, and real in the same way, as the gases the laws of which it explains.   It is an idea essential to the intelligibility of the world not to one mind, but to all ; it is an idea which nature as well as mankind accepts. That, I maintain, is the test and the very meaning of reality.

CHAPTER VI

MEASUREMENT

WE have now examined the chief types of scientific proposition and discussed what are the principles and the facts on which all science rests. Already we have had occasion to notice differences between the various branches of science, and when we leave such very fundamental questions the differences are bound to become more prominent. It does not seem to me that there is much left to say (except in amplification of what has been said already) that would be applicable to the whole of science. But there is one further matter which may fitly receive some attention; for though it affects only part of science, that part is constantly growing both in volume and in importance. Moreover, the sciences into which it enters are generally held to be particularly difficult to popularize, and to be beyond the reach of the untrained reader. Accordingly, we shall scarcely diverge from the main purpose of our discussion if in the next two chapters we give it some consideration.

This matter is measurement and all the structure of mathematical science which rests upon measurement. Every one knows that measurement is a very important part of many sciences; they know too, that many sciences are " mathematical " and can only be apprehended completely by those versed in mathematics. But very few people could explain exactly how measurement enters into science, why it enters into some and not in others, why it is so important, what mathematics is and why it is so intimately connected with measurement and with the sciences in which measurement is involved. In the next two chapters I propose to attempt some answer to these questions. Any answer to these questions that can be attempted here will not, of course, enable anyone to start immediately the study of one of the mathematical sciences in the hope of understanding it completely. But if he can be convinced that even in the most abstruse parts of those sciences there is something that he can comprehend and appreciate without the smallest knowledge of mathematics, something may be done towards extending the range of the sciences that are open to the layman.

WHAT IS MEASUREMENT ?

Measurement is one of the notions which modern science has taken over from common sense. Measurement does not appear as part of common sense until a comparatively high stage of civilization is reached ; and even the common-sense conception has changed and developed enormously in historic times. When I say that measurement belongs to common sense, I only mean that it is something with which every civilized person to-day is entirely familiar. It may be defined, in general, as the assignment of numbers to represent properties. If we say that the time is 3 o'clock, that the price of coal is 56 shillings a ton, and that we have just bought 2 tons of it—in all such cases we are using numbers to convey important information about the " properties " of the day, of coal in general, of the coal in our cellar, or so on ; and our statement depends somehow upon measurement.

The first point I want to notice is that it is only some properties and not all that can be thus represented by numbers. If I am buying a sack of potatoes I may ask what it weighs and what it costs ; to those questions I shall expect a number in answer ; it weighs 56 lbs.

and costs 5 shillings. But I may also ask of what variety the potatoes are, and whether they are good cookers ; to those questions I shall not expect a number in answer. The dealer may possibly call the variety " No. 11 " in somebody's catalogue ; but even if he does, I shall feel that such use of a number is not real measurement, and is not of the same kind as the use in connexion with weight or cost. What is the difference ? Why are some properties measurable and others not ? Those are the questions I want to discuss. And I will outline the answer immediately in order that the reader may see at what the subsequent discussion is aiming. The difference is this. Suppose I have two sacks of potatoes which are identical in weight, cost, variety, and cooking qualities ; and that I pour the two sacks into one so that there is now only one sack of potatoes. This sack will differ from the two original sacks in weight and cost (the measurable properties), but will not differ from them in variety and cooking qualities (the properties that are not measurable). The measurable properties of a body are those which are changed by the combination of similar bodies ; the non-measurable properties are those that are not changed. We shall see that this definition is rather too crude, but it will serve for the present.

NUMBERS

In order to see why this difference is so important we must inquire more closely into the meaning of " number." And at the outset we must note that confusion is apt to arise because that word is used to denote two perfectly different things. It sometimes means a mere name or word or symbol, and it sometimes means a property of an object. Thus, besides the properties which have been mentioned, the sack of potatoes has another definite property, namely the number of potatoes in it, and the number is as much a property of the object which we call a sack of potatoes as its weight or its cost. This property can be (and must be) " represented by a number " just as the weight can be ; for instance, it might be represented by 200. But this " 200 " is not itself a property of the sack ; it is a mere mark on the paper for which would be substituted, if I was speaking instead of writing, a spoken sound ; it is a name or symbol for the property. When we say that measurement is the representation of properties by " numbers," we mean that it is the representation of properties, other than number, by the symbols which are always used to represent number. Moreover, there is a separate word for these symbols ; they are called " numerals." We shall always use that word in future and confine " number " to the meaning of the property which is always represented by numerals.

These considerations are not mere quibbling over words ; they bring out clearly an important point, namely, that the measurable properties of an object must resemble in some special way the property number, since they can be fitly represented by the same symbols ; they must have some quality common with number. We must proceed to ask what this common quality is, and the best way to begin is to examine the property number rather more closely.

The number of a sack of potatoes, or, as it is more usually expressed, the number of potatoes contained in it, is ascertained by the process of counting. Counting is inseparably connected in our minds to-day with numerals, but the process can be, and at an earlier stage of civilization was, carried on without them. Without any use of numerals I can determine whether the number of one sack of potatoes is equal to that of another. For this purpose I take a potato from one sack, mark it in some way to distinguish it from the rest (e.g. by putting it into a box), and then perform a similar operation on a potato from the other sack. I then repeat this double operation continually until I have exhausted the potatoes from one sack. If the operation which exhausts the potatoes from one sack exhausts also the potatoes from the other, then I know that the sacks had the same number of potatoes ; if not, then the sack which is not exhausted had a larger number of potatoes than the other.

THE RULES FOR COUNTING

This process could be applied equally well if the objects counted against each other were not of the same nature. The potatoes in a sack can be counted, not only against another collection of potatoes, but also against the men in a regiment or against the days in the year. The " mark," which is used for distinguishing the objects in the process of counting, may have to be altered to suit the objects counted, but some other suitable mark could be found which would enable the process to be carried out. If, having never heard of counting before, we applied the process to all kinds of different objects, we should soon discover certain rules which would enable us to abbreviate and simplify the process considerably. These rules appear to us to-day so obvious as to be hardly worth stating, but as they are undoubtedly employed in modern methods of counting, we must notice them here. The first is that if two sets of objects, when counted against a third set, are found to have the same number as that third set, then, when counted against each other they will be found to have the same number. This rule enables us to determine whether two sets of objects have the same number without bringing them together ; if I want to find out whether the number of potatoes in the sack I propose to buy is the same as that in a sack I have at home, I need not bring my sack to the shop ; I can count the potatoes at the shop against some third collection, take this collection home, and count it against my potatoes. Accordingly the discovery of this first rule immediately suggests the use of portable collections which can be counted, first against one collection and then against another, in order to ascertain whether these two have the same number.

The value of this suggestion is increased greatly by the discovery of a second rule. It is that by starting with a single object and continually adding to it another single object, we can build up a series of collections of which one will have the same number as any other collection whatsoever. This rule helps us in two ways. First, since it states that it is possible to make a standard series of collections one of which will have the same number as any other collection, it suggests that it might be well to count collections, not against each other, but against a standard series of collections. If we could carry this standard series about with us, we could always ascertain whether any one collection had the same number as any other by observing whether the member of the standard series which had the same number as the first had also the same number as the second. Next, it shows us how to make such a standard series with the least possible cumbrousness. If we had to have a totally different collection for each member of the standard series, the whole series would be impossibly cumbrous ; but our rule shows that the earlier members of the series (that is those with the smaller number) may be all parts of the later members. Suppose we have a collection of objects, each distinguishable from each other, and agree to take one of these objects as the first member of the series ; this object together with some other as the next member ; these objects with yet another as the next member ; and so on. Then we shall obtain, according to our rule, a series, some member of which has the same number as any other collection we want to count, and yet the number of objects, in all the members of the standard series taken together, will not be greater than that of the largest collection we want to count.

And, of course, this is the process actually adopted. For the successive members of the standard series compounded in this way, primitive man chose, as portable, distinguishable objects, his fingers and toes. Civilized man invented numerals for the same purpose. Numerals are simply distinguishable objects out of which we build our standard series of collections by adding them in turn to previous members of the series. The first member of our standard series is 1, the next 1, 2, the next 1, 2, 3 and so on. We count other collections against these members of the standard series and so ascertain whether or no two collections so counted have the same number. By an ingenious convention we describe which member of the series has the same number as a collection counted against it by quoting simply the last numeral in that member ; we describe the fact that the collection of the days of the week has the same number as the collection 1, 2, 3, 4, 5, 6, 7, by saying " that the number " of the days of the week is 7. But when we say that what we really mean, and what is really important, is that this collection has the same number as the collection of numerals (taken in the standard order) which ends in 7 and the same number as any other collection which also has the same number as the collection of numerals which ends in 7.[1]

---

[1] Numerals have also an immense advantage over fingers and toes as objects of which the standard series may be formed, in that the series can be extended indefinitely by a simple rule which automatically gives names to any new numerals that may be required. Even if we have never hitherto had reason to carry the series beyond (say) 131679 in order to count all the collections we have met with, when we do meet at last with a larger collection, we know at once that the objects

The two rules that have been mentioned are necessary to explain what we mean by " the number " of a collection and how we ascertain that number. There is a third rule which is of great importance in the use of numbers. We often want to know what is the number of a collection which is formed by combining two other collections of which the numbers are known, or, as it is usually called, adding the two collections. For instance we may ask what is the number of the collection made by adding a collection of 2 objects to a collection of 3 objects. We all know the answer, 5. It can be found by arguing thus : The first collection can be counted against the numerals 1, 2 ; the second against the numerals 1, 2, 3. But the numerals 1, 2, 3, 1, 2, a collection formed by adding the two first collections, can be counted against 1, 2, 3, 4, 5. Therefore the number of the combined collection is 5. However, a little examination will show that in reaching this conclusion we have made use of another rule, namely that if two collections A and a, have the same number, and two other collections B and b, have the same number, then the collection formed by adding A to B has the same number as that formed by adding a to b ; in other words, equals added to equals produce equal sums. This is a third rule about numbers and counting ; it is quite as important as the other two rules ; all three are so obvious to us to-day that we never think about them, but they must have been definitely discovered at some time in the history of mankind, and without them all, our habitual use of numbers would be impossible.

we must add to our standard series are 131680, 131681, and so on. This is a triumph of conventional nomenclature, much more satisfactory than the old convention that when we have exhausted our fingers we must begin on our toes, but it is not essentially different.

And now, after this discussion of number, we can return to the other measurable properties of objects which, like number, can be represented by numerals. We can now say more definitely what is the characteristic of these properties which makes them measurable. It is that there are rules true of these properties, closely analogous to the rules on which the use of number depends. If a property is to be measurable it must be such that (1) two objects which are the same in respect of that property as some third object are the same as each other ; (2) by adding objects successively we must be able to make a standard series one member of which will be the same in respect of the property as any other object we want to measure ; (3) equals added to equals produce equal sums. In order to make a property measurable we must find some method of judging equality and of adding objects, such that these rules are true.

Let me explain what is meant by using as an example the measurable property, weight.

Weight is measured by the balance. Two bodies are judged to have the same weight if, when they are placed in opposite pans, neither tends to sink ; and two bodies are added in respect of weight when they are both placed on the same pan of the balance. With these definitions of equality and addition, it is found that the three rules are obeyed. (1) If the body A balances the body B, and B balances C, then A balances C. (2) By placing a body in one pan and continually adding it to others, collections can be built up which will balance any other body placed in the other pan.[1] (3) If the body A balances the body B, and C balances D, then A and C in the same pan will balance B and D in the other pan. To make the matter yet clearer let us take another measurable property,

[1] See further, p. 122.

length.   Two straight rods are judged equal in length, if they can be placed so that both ends of one are contiguous to both ends of the other ; they are added in respect of length, when they are placed with one end of one contiguous with one end of the other, while the two form a single straight rod.   Here again we find the three rules fulfilled.   Bodies which are equal in length to the same body are equal in length to each other.   By adding successively rods to each other, a rod can be built up which is equal to any other rod.   And equal rods added to equal rods produce equal rods.   Length is therefore a measurable property.

It is because these rules are true that measurement of these properties is useful and possible ; it is these rules that make the measurable properties so similar to numbers, that it is possible and useful to represent them by numerals the primary purpose of which is to represent numbers.   It is because of them that it is possible to find one, and only one numeral, which will fitly represent each property ; and it is because of them, that these numerals, when they are found, tell us something useful about the properties.   One such use arises in the combination of bodies possessing the properties.   We may want to know how the property varies when bodies possessing it are added in the way characteristic of measurement. When we have assigned numerals to represent the property we shall know that the body with the property 2 added to that with the property 3 will have the same property as that with the property 5, or as the combination of the bodies with properties 4 and 1.   This is not the place to examine exactly how these conclusions are shown to be universally valid ; but they are valid only because the three rules are true.

But what is the nature of these rules ? They are laws established by definite experiment. The word " rule" has been used hitherto, because it is not quite certain whether they are truly laws in their application to number ; but they certainly are laws in their application to other measurable properties, such as weight or length. The fact that the rules are true can be, and must be, determined by experiment in the same way as the fact that any other laws are true. Perhaps it may have appeared to the reader that the rules must be true ; that it requires no experiment to determine that bodies which balance the same body will balance each other ; and that it is inconceivable that this rule should not be true. But I think he will change his opinion, if it is pointed out that the rule is actually true only in certain conditions ; for instance, it is only true if the balance is a good one, and has arms of equal length and pans of equal weight. If the arms were unequal, the rule would not be found to be true unless it were carefully prescribed in which pan the bodies were placed during the judgment of equality. Again, the rules would not be true of the property length, unless the rods were straight and were rigid. In implying that the balance is good, and the rods straight and rigid, we have implied definite laws which must be true if the properties are to be measurable, namely that it is possible to make a perfect balance, and that there are rods which are straight and rigid. These are experimental laws ; they could not be known apart from definite experiment and observation of the external world ; they are not self-evident.

Accordingly the process of discovering that a property is measurable in the way that has been described, and setting up a process for measuring it, is one that rests entirely upon experimental inquiry. It is a part, and a most important part, of experimental science. Whenever a new branch of physics is opened up (for, as has been said, physics is the science that deals with such processes of measurement), the first step is always to find some process for measuring the new properties that are investigated ; and it is not until this problem has been solved, that any great progress can be made along the branch. Its solution demands the discovery of new laws. We can actually trace the development of new measurable properties in this way in the history of science. Before the dawn of definite history, laws had been discovered which made measurable some of the properties employed by modern science. History practically begins with the Greeks, but before their time the properties, weight, length, volume, and area had been found to be measurable ; the establishment of the necessary laws had probably occurred in the great period of Babylonian and and Egyptian civilization. The Greeks, largely in the person of Archimedes, found how to measure force by establishing the laws of the lever, and other mechanical systems. Again from the earliest era, there have been rough methods of measuring periods of time,[1] but a true method, really obeying the three rules, was not discovered till the seventeenth century ; it arose out of Galileo's laws of the pendulum. Modern science has added greatly to the list of measurable properties ; the science of electricity is based on the discovery, by Cavendish and Coulomb, of the law necessary to measure an electric charge ; on the laws, discovered by Œrsted and Ampère, necessary to measure an electric current ; and on the laws, discovered by Ohm and Kirchhoff, necessary to

---

[1] By a period of time I mean the thing that is measured when we say that it took us 3 hours to do so-and-so. This is a different " time " from that which is measured when we say it is 3 o'clock. The difference is rather abstruse and cannot be discussed here ; but it may be mentioned that the " measurement " involved in " 3 o'clock " is more like that discussed later in the chapter.

measure electrical resistance. And the discovery of similar laws has made possible the development of other branches of physics.

But, it may be asked, has there ever been a failure to discover the necessary laws ? The answer is that there are certainly many properties which are not measurable in the sense that we have been discussing ; there are more properties, definitely recognized by science, that are not so measurable than are so measurable. But, as will appear presently, the very nature of these properties makes it impossible that they should be measured in this way. For the only properties to which this kind of measurement seems conceivably applicable, are those which fulfil the condition stated provisionally on p. 111 ; they must be such that the combination of objects possessing the property increases that property. For this is the fundamental significance of the property number ; it is something that is increased by addition ; any property which does not agree with number in this matter cannot be very closely related to number and cannot possibly be measured by the scheme that has been described. But it will be seen that fulfilment of this condition only makes rule (2) true ; it is at least conceivable that a property might obey rule (2) and not rules (1) and (3). Does that ever happen, or can we always find methods of addition and of judging equality such that, if rule (2) is true, the laws are such that rules (1) and (3) are also true ? In the vast majority of cases we can find such methods and such laws ; and it is a very remarkable fact that we can ; it is only one more instance of the way in which nature kindly falls in with our ideas of what ought to be. But I think there is one case in which the necessary methods and laws have not yet been found and are not likely to be found. It is a very difficult matter concerning which even expert physicists might differ, and so no discussion of it can be entered on here. But it is mentioned in order to impress the reader with the fact that measurement does depend upon experimental laws ; that it does depend upon the facts of the external world ; and that it is not wholly within our power to determine whether we will or will not measure a certain property.   That is the feature of measurement which it is really important to grasp for a proper understanding of science.

MULTIPLICATION

Before we pass to another kind of measurement reference must be made to a matter which space does not allow to be discussed completely.   In stating the rules that were necessary in order that weight should be measurable (p. 117), it was said that a collection having the same weight as any given body could be made by adding other bodies to that first selected.   Now this statement is not strictly true ; it is only true if the body first selected has a smaller weight than any other body it is desired to weigh ; and even if this condition is fulfilled, it is not true if the bodies added successively to the collection are of the same weight as that first selected.   Thus if my first body weighs 1 lb., I cannot by adding to it make a collection which weighs less than 1 lb., and by adding bodies which each weigh 1 lb., I cannot make a collection which has the same weight as a body weighing (say) $2\frac{1}{2}$ lb.

These facts, to which there is no true analogy in connexion with number, force us to recognize " fractions." A considerable complication is thereby introduced, and the reader must accept my assurance that they can all be solved by simple developments of the process of measurement that has been sketched.   But for a future purpose it is necessary to notice very briefly the processes of the multiplication and division of magnitudes on which the significance of fractions depends.

Suppose I have a collection of bodies, each of which has the same weight 3, the number of bodies in the collection being 4. I may ask what is the weight of the whole collection. The answer is given of course by multiplying 3 by 4, and we all know now that the result of that operation is 12. That fact, and all the other facts summed up in the multiplication table which we learn at school, can be proved from the rules on which weighing depend together with facts determined by counting numerals. But the point I want to make is that multiplication represents a definite experimental operation, namely the combination into a single collection, placed on one pan of the balance, of a set of bodies, all of the same weight, the number of those bodies being known. Division arises directly out of multiplication. In place of asking what will be the weight of a collection formed of a given number of bodies all of the same weight, we ask what must be the weight of each of a collection of bodies, having a given number, when the whole collection has a given weight. E.g. what must each body weigh in order that the whole collection of 4 bodies weighs 12 ? The answer is obtained by dividing 12 by 4. That answer is obtained, partly from the multiplication table, partly by inventing new numerals which we call fractions ; but once again division corresponds to a definite experimental operation and has its primary significance because it corresponds to that operation. This is this conclusion that we shall use in the sequel. But it is worth while noting that the fractions which we obtain by this method of addition overcome the difficulty from which this paragraph started. If we make all possible fractions of our original weight (i.e. all possible bodies, such that some number of them formed into a single collection have the same weight as the original body), then, by adding together suitable collections of these fractions, we can make up a collection which will have the same weight as any body whatever that we desire to weigh. This result is an experimental fact which could not have been predicted without experimental inquiry. And the result is true, not only for the measurable property weight, but for all properties measurable by the process that is applicable to weight. Once more we see how much simpler and more conveniently things turn out than we have really any right to expect ; measurement would have been a much more complex business if the law that has just been stated were not always true.

DERIVED MEASUREMENT

Measurement, it was said on p. 110, is the assignment of numbers (or, as we say now, numerals) to represent properties. We have now considered one way in which this assignment is made, and have brought to light the laws which must be true if this way is to be possible. And it is the fundamental way. We are now going to consider some other ways in which numerals are assigned to represent properties ; but it is important to insist at the outset, and to remember throughout, that these other ways are wholly dependent upon the fundamental process, which we have just been discussing, and must be so dependent if the numerals are to represent " real properties " and to tell us something scientifically significant about the bodies to which they are attached. This statement is confirmed by history ; all properties measured in the definitely pre-scientific era were measured (or at least measurable) by the fundamental process ; that is true of weight, length, volume, area and periods of time. The dependent measurement, which we are now about to consider, is a product of definitely and consciously scientific investigation, although the actual discovery may, in a few cases, be lost in the mists of the past.

The property which we shall take as an example of this dependent or, as it will be termed, derived measurement, is *density*. Every one has some idea of what density means and realizes, vaguely at least, why we say that iron is denser than wood or mercury than water ; and most people probably know how density is measured, and what is meant when it is said that the density of iron is 8 times that of wood, and the density of mercury $13\frac{1}{2}$ times that of water. But they will feel also that there is something more scientific and less purely common-sense about the measurement of density than about the measurement of weight ; as a matter of fact the discovery of the measurement of density certainly falls within the historic period and probably may be attributed to Archimedes (about 250 B.C.). And a little reflection will convince them that there is something essentially different in the two processes.

For what we mean when we say a body has a weight 2 is that a body of the same weight can be made by combining 2 bodies of the weight 1 ; that is the fundamental meaning of weight ; it is what makes weight physically important and, as we have just seen, makes it measurable. But when we say that mercury has a density $13\frac{1}{2}$ we do *not* mean that a body of the same density can be prepared by combining $13\frac{1}{2}$ bodies of the density 1 (water). For, if we did mean that, the statement would not be true. However many pieces of water we take, all of the same density, we cannot produce a body with any different density. Combine water with water as we will, the resulting body has the density of water. And this, a little reflection will show, is part of the fundamental meaning of density ; density is something that is characteristic of all pieces of water, large and small. The density of water, a " quality " of it, is something fundamentally independent of and in contrast with the weight of water, the " quantity " of it.

But the feature of density, from which it derives its importance, makes it totally impossible to measure density by the fundamental process discussed earlier in the chapter. How then do we measure it? Before we answer that question, it will be well to put another. As was insisted before, if measurement is really to mean anything, there must be some important resemblance between the property measured, on the one hand, and the numerals assigned to represent it, on the other. In fundamental measurement, this resemblance (or the most important part of it) arises from the fact that the property is susceptible to addition following the same rules as that of number, with which numerals are so closely associated. That resemblance fails here. What resemblance is left?

MEASUREMENT AND ORDER

There is left a resemblance in respect of " order." The numerals are characterized, in virtue of their use to represent numbers, by a definite order; they are conventionally arranged in a series in which the sequence is determined : " 2 " follows " 1 " and is before " 3 "; " 3 " follows " 2 " and is before " 4 " and so on. This characteristic order of numerals is applied usefully for many purposes in modern life; we " number " the pages of a book or the houses of a street, not in order to know the number of pages in the book or of houses in the street—nobody but the printer or the rate-surveyor cares about that—but in order to be able to find any given page or house easily. If we want p. 201 and the book opens casually at p. 153 we know in which direction to turn the pages.[1] Order then is characteristic

---

[1] Numerals are also used to represent objects, such as soldiers or telephones, which have no natural order. They are used here because they provide an *inexhaustible* series of names, in virtue of the ingenious device by which new numerals can always be invented when the old ones have been used up.

of numerals ; it is also characteristic of the properties represented by numerals in the manner we are considering now. This is our feature which makes the " measurement " significant. Thus, in our example, bodies have a natural order of density which is independent of actual measurement. We might define the words " denser " or " less dense " as applied to liquids (and the definition could easily be extended to solids) by saying that the liquid A is denser than B, and B less dense than A, if a substance can be found which will float in A but not in B. And, if we made the attempt, we should find that by use of this definition we could place all liquids in a definite order, such that each member of the series was denser than the preceding and less dense than the following member. We might then assign to the first liquid the density 1, to the second 2, and so on ; and we should then have assigned numerals in a way which would be physically significant and indicate definite physical facts. The fact that A was represented by 2 and B by 7 would mean that there was some solid body which would float in B, but not in A. We should have achieved something that might fairly be called measurement.

Here again it is important to notice that the possibility of such measurement depends upon definite laws ; we could not have predicted beforehand that such an arrangement of liquids was possible unless we knew these laws. One law involved is this : If B is denser than A, and C denser than B, then C is denser than A. That sounds like a truism ; but it is not. According to our definition it implies that the following statement is always true : If a body X floats in B and sinks in A, then if another body Y sinks in B it will also sink in A. That is a statement of facts ; nothing but experiment could prove that it is true ; it is a law. And if it were not true, we could not arrange liquids naturally in a definite order. For the test with X would prove that B was denser than A, while the test with Y (floating in A, but sinking in B) would prove that A was denser than B. Are we then to put A before or after B in the order of density ? We should not know. The order would be indeterminate and, whether we assigned a higher or a lower numeral to A than to B, the assignment would represent no definite physical fact : it would be arbitrary.

In order to show that the difficulty might occur, and that it is an experimental law that it does not occur, an instance in which a similar difficulty has actually occurred may be quoted. An attempt has been made to define the " hardness " of a body by saying that A is harder than B if A will scratch B. Thus diamond will scratch glass, glass iron, iron lead, lead chalk, and chalk butter ; so that the definition leads to the order of hardness : diamond, glass, iron, lead, chalk, butter. But if there is to be a definite order, it must be true in all cases that if A is harder than B and B than C, then A is harder than C ; in other words, if A will scratch B and B C, then A will scratch C. But it is found experimentally that there are exceptions to this rule, when we try to include all bodies within it and not only such simple examples as have been quoted. Accordingly the definition does not lead to a definite order of hardness and does not permit the measurement of hardness.

There are other laws of the same kind that have to be true if the order is to be definite and the measurement significant ; but they will not be given in detail. One of them the reader may discover for himself, if he will consider the property colour. Colour is not a property measurable in the way we are considering, and for this reason. If we take all reds (say) of a given shade, we can arrange them definitely in an order of lightness and darkness ; but no colour other than red will fall in this order. On the other hand, we might possibly take all shades and arrange them in order of redness—pure red, orange, yellow, and so on ; but in this order there would be no room for reds of different lightness. Colours cannot be arranged in a single order, and it is for this reason that colour is not measurable as is density.

NUMERICAL LAWS

But though arrangement in this manner in an order and the assignment of numerals in the order of the properties are to some extent measurement and represent something physically significant, there is still a large arbitrary element involved. If the properties A, B, C, D, are naturally arranged in that order, then in assigning numerals to represent the properties I must *not* assign to A 10, to B 3, to C 25, to D 18 ; for if I did so the order of the numerals would not be that of the properties. But I have an endless number of alternatives left ; I might put A 1, B 2, C 3, D 4 ; or A 10, B 100, C 1,000, D 10,000 ; or A 3, B 9, C 27, D 81 ; and so on. In the true and fundamental measurement of the first part of the chapter there was no such latitude. When I had fixed the numeral to be assigned to one property, there was no choice at all of the numerals to be assigned to the others ; they were all fixed. Can I remove this latitude here too and find a way of fixing definitely what numeral is to be assigned to represent each property ?

In some cases, I can ; and one of these cases is density. The procedure is this. I find that by combining the numerals representing other properties of the bodies, which can be measured definitely according to the fundamental process, I can obtain a numeral for each body, and that these numerals lie in the order of the property I want to measure. If I take these numerals as representing the property, then I still get numerals in the right order, but the numeral for each property is definitely fixed. An example will be clearer than this general statement. In

the case of density, I find that if I measure the weight
and the volume of a body (both measurable by the fundamental process and therefore definitely fixed), and I divide
the weight by the volume, then the numerals thus
obtained for different bodies lie in the order of their densities, as density was defined on p. 127. Thus I find that
1 gallon of water weighs 10 lb., but 1 gallon of mercury
weighs 135 lb. ; the weight divided by the volume for
water is 10, for mercury is 135 ; 135 is greater than 10 ;
accordingly, if the method is correct, mercury should be
denser than water and any body which sinks in mercury
should sink in water. And that is actually found to be
true. If therefore I take as the measure of the density
of a substance, its weight divided by its volume, then I
get a number which is definitely fixed,[1] and the order of
which represents the order of density. I have arrived
at a method of measurement which is as definitely fixed
as the fundamental process and yet conveys adequately
the physically significant facts about order.

The invention of this process of measurement for
properties not suited for fundamental measurement is a
very notable achievement of deliberate scientific investigation. The process was not invented by common
sense ; it was certainly invented in the historic period,
but it was not until the middle of the eighteenth century
that its use became widespread.[2] To-day it is one of the
most powerful weapons of scientific investigation ; and
it is because so many of the properties of importance to

---

[1] Except in so far as I may change the units in which I measure
weights and volume. I should get a different number if I measured
the volume in pints and the weight in tons. But this latitude in the
choice of units introduces a complication which it will be better to
leave out of account here. There is no reason why we should not
agree once and for all to use the same units ; and if we did that the
complication would not arise.

[2] I think that until the eighteenth century only two properties were
measured in this way which were not measurable by the fundamental
process, namely density and constant acceleration.

other sciences are measured in this way that physics, the science to which this process belongs, is so largely the basis of other sciences. But it may appear exceedingly obvious to the reader, and he may wonder why the invention was delayed so long. He may say that the notion of density, in the sense that a given volume of the denser substance weighs more than the same volume of the less dense, is the fundamental notion ; it is what we mean when we speak of one substance being denser (or in popular language " heavier ") than another ; and that all that has been discovered in this instance is that the denser body, in this sense, is also denser in the sense of p. 127. This in itself would be a very noteworthy discovery, but the reader who raises such an objection has overlooked a yet more noteworthy discovery that is involved.

For we have observed that it is one of the most characteristic features of density that it is the same for all bodies, large and small, made of the same substance. It is this feature which makes it impossible to measure it by the fundamental process. The new process will be satisfactory only if it preserves this feature. If we are going to represent density by the weight divided by the volume, the density of all bodies made of the same substance will be the same, as it should be, only if for all of them the weight divided by the density is the same, that is to say, in rather more technical language, if the weight is proportional to the density. In adopting the new process for measuring density and assigning numerals to represent it in a significant manner, we are, in fact, assuming that, for portions of the same substance, whether they are large or small, the weight is proportional to the volume. If we take a larger portion of the same substance and thereby double the weight, we must find, if the process of measurement is to be a success, that we also double the volume ; and this law must be true for all substances to which the conception of density is applicable at all.

Of course every one knows that this relation is actually true ; it is so familiar that we are apt to forget that it is an experimental truth that was discovered relatively late in the history of civilization, which easily might not be true. Perhaps it is difficult to-day to conceive that when we take " more " of a substance (meaning thereby a greater volume) the weight should not increase, but it is quite easy to conceive that the weight should not increase proportionally to the volume ; and yet it is upon strict proportionality that the measurement of density actually depends. If the weight had not been proportional to the volume, it might still have been possible to measure density, so long as there was some fixed numerical relation between weight and volume. It is this idea of a fixed numerical relation, or, as we shall call it henceforward, a numerical law, that is the basis of the " derived " process of measurement that we are considering ; and the process is of such importance to science because it is so intimately connected with such numerical laws. The recognition of such laws is the foundation of modern physics.

THE IMPORTANCE OF MEASUREMENT

For why is the process of measurement of such vital importance ; why are we so concerned to assign numerals to represent properties. One reason doubtless is that such assignment enables us to distinguish easily and minutely between different but similar properties. It enables us to distinguish between the density of lead and iron far more simply and accurately than we could do by saying that lead is rather denser than iron, but not nearly so dense as gold—and so on. But for that purpose the " arbitrary " measurement of density, depending simply on the arrangements of the substances in their order (p. 127), would serve equally well. The true answer to our question is seen by remembering the conclusion, at which we arrived in Chapter III, that the terms between which laws express relationships are themselves based on laws and represent collections of other terms related by laws. When we measure a property, either by the fundamental process or by the derived process, the numeral which we assign to represent it is assigned as the result of experimental laws ; the assignment implies laws. And therefore, in accordance with our principle, we should expect to find that other laws could be discovered relating the numerals so assigned to each other or to something else ; while if we assigned numerals arbitrarily without reference to laws and implying no laws, then we should not find other laws involving these numerals. This expectation is abundantly fulfilled, and nowhere is there a clearer example of the fact that the terms involved in laws themselves imply laws. When we can measure a property truly, as we can volume (by the fundamental process) or density (by the derived process) then we are always able to find laws in which these properties are involved ; we find, e.g., the law that volume is proportional to weight or that density determines, in a certain precise fashion, the sinking or floating of bodies. But when we cannot measure it truly, then we do not find a law. An example is provided by the property " hardness " (p. 128) ; the difficulties met with in arranging bodies in order of hardness have been overcome ; but we still do not know of any way of measuring, by the derived process, the property hardness ; we know of no numerical law which leads to a numeral which always follows the order of hardness. And so, as we expect, we do not know any accurate and general laws relating hardness to other properties. It is because true measurement is essential to the discovery of laws that it is of such vital importance to science.

One final remark should be made before we pass on. In this chapter there has been much insistence on the distinction between fundamental measurement (such as is applicable to weight) and derived measurement (such as is applicable to density). And the distinction is supremely important, because it is the first kind of measurement which alone makes the second possible. But the reader who, when he studies some science in detail, tries, as he should, to discover which of the two processes is involved in the measurement of the various properties characteristic of that science, may occasionally find difficulty in answering the question. It should be pointed out, therefore, that it is quite possible for a property to be measurable by both processes. For all properties measurable by the fundamental process must have a definite order ; for the physical property, number, to which they are so similar, has an order—the order of " more " or " less." This order of number is reflected in the order of the numerals used to represent number. But if it is to be measurable by the derived process, it must also be such that it is also a " constant " in a numerical law—a term that is just going to be explained in the next chapter. There is nothing in the nature of fundamental measurement to show that a property to which it is applicable may not fulfil this condition also ; and sometimes the condition is fulfilled, and then the property is measurable either by the fundamental or the derived process. However, it must be remembered that the properties involved in the numerical law must be such that they are fundamentally measurable ; for otherwise the law could not be established. The neglect of this condition is apt to lead to confusion ; but with this bare hint the matter must be left.

CHAPTER VII

NUMERICAL LAWS AND THE USE OF MATHEMATICS IN SCIENCE

NUMERICAL LAWS

IN the previous chapter we concluded that density was a measurable property because there is a fixed numerical relation, asserted by a " numerical law," between the weight of a substance and its volume. In this chapter we shall examine more closely the idea of a numerical law, and discover how it leads to such exceedingly important developments.

Let us first ask exactly what we do when we are trying to discover a numerical law, such as that between weight and volume. We take various portions of a substance, measure their weights and their volumes, and put down the result in two parallel columns in our notebook. Thus I may find these results :

TABLE I

WEIGHT	VOLUME	WEIGHT	VOLUME
1	7	4	28
2	14	10	70
3	21	29	203

I now try to find some fixed relation between the corresponding numbers in the two columns ; and I shall succeed in that attempt if I can find some rule whereby, starting with the number in one column, I can arrive at the corresponding number in the other. If I find such a rule— and if the rule holds good for all the further measurements that I may make—then I have discovered a numerical law.

In the example we have taken the rule is easy to find.

I have only to divide the numbers in the second column
by 7 in order to arrive at those in the first, or multiply
those in the first by 7 in order to arrive at those in the
second.   That is a definite rule which I can always apply
whatever the numbers are ; it is a rule which might always
be true, but need not always be true ; whether or no it
is true is a matter for experiment to decide.   So much is
obvious ; but now I want to ask a further and important
question.   How did we ever come to discover this rule ;
what suggested to us to try division or multiplication
by 7 : and what is the precise significance of division and
multiplication in this connexion ?

THE SOURCE OF NUMERICAL RELATIONS

The answer to the first part of this question is given by
the discussion on p. 123.   Division and multiplication
are operations of importance in the counting of objects ;
in such counting the relation between 21, 7, 3 (the third
of which results from the division of the first by the
second) corresponds to a definite relation between the
things counted ; it implies that if I divide the 21 objects
into 7 groups, each containing the same number of objects,
then the number of objects in each of the 7 groups is 3.
By examining such relations through the experimental
process of counting we arrive at the multiplication (or
division) table.   This table, when it is completed, states
a long series of relations between numerals, each of which
corresponds to an experimental fact ; the numerals
represent physical properties (numbers) and in any given
relation (e.g. $7 \times 3 = 21$) each numeral represents a
different property.   But when we have got the multiplication table, a statement of relations between numerals,
we can regard it, and do usually regard it, *simply* as a
statement of relations between numerals ; we can think
about it without any regard to what those numerals

represented when we were drawing up the table. And if any other numerals are presented to our notice, it is possible and legitimate to ask whether these numerals, whatever they may represent, are in fact related as are the numerals in the multiplication table. In particular, when we are seeking a numerical relation between the columns of Table I, we may inquire, and it is natural for us to inquire, whether by means of the multiplication we can find a rule which will enable us to arrive at the numeral in the second column starting from that in the first.

That explains why it is so natural to us to try division when we are seeking a relation between numbers. But it does not answer the second part of the question ; for in the numerical law that we are considering, the relation between the things represented by the numerals is *not* that which we have just noticed between things counted. When we say that, by dividing the volume by 7, we can arrive at the weight, we do *not* mean that the weight *is* the volume of each of the things at which we arrive by dividing the substance into 7 portions, each having the same volume. For a weight can never *be* a volume, any more than a soldier can *be* a number ; it can only be represented by the same numeral as a volume, as a soldier can be represented by a numeral which also represents a number.

The distinction is rather subtle, but if the reader is to understand what follows, he must grasp it. The relation which we have found between weight and volume is a pure numerical relation ; it is suggested by the relation between actual things, namely collections which we count ; but it is not that relation. The difference may be expressed again by means of the distinction between numbers and numerals. The relation between actual things counted is a relation between the numbers—which are physical properties—of those things ; the relation

between weight and volume is a relation between numerals,
the numerals that are used to represent those properties.
The physical relation in the second case is not between
numbers at all, but between weight and volume which
are properties quite different from numbers ; it appears
very similar to that between numbers only because we
use numerals, originally invented to represent numbers,
to represent other properties.   The relation stated by a
numerical law is a relation between numerals, and only
between numerals, though the idea that there may be
such a relation has been suggested to us by the study of
the physical property, number.

If we understand this, we shall see what a very remarkable thing it is that there should be numerical laws at all,
and shall see why the idea of such a law arose comparatively late in the history of science.   For even when we
know the relations between numbers, there is no reason
to believe that there must be any relations of the same
kind between the numerals which are used to represent,
not only numbers, but also other properties.   Until we
actually tried, there was no reason to think that it must
be possible to find at all numerical laws, stating numerical
relations such as those of division and multiplication.
The fact that there are such relations is a new fact, and
ought to be surprising.   As has been said so often, it
does frequently turn out that suggestions made simply
by our habits of mind are actually true ; and it is because
they are so often true that science is interesting.   But
every time they are true there is reason for wonder and
astonishment.

And there is a further consequence yet more deserving
of our attention at present   If we realize that the
numerical relations in numerical laws, though suggested
by relations between numbers, are not those relations,
we shall be prepared to find also numerical relations which
are not even suggested by relations between numbers,

but only by relations between numerals. Let me take an example. Consider the pairs of numerals (1, 1), (2, 4), (3, 9), (4, 16) . . . Our present familiarity with numerals enables us to see at once what is the relation between each pair ; it is that the second numeral of the pair is arrived at by multiplying the first numeral *by itself* ; 1 is equal to 1 × 1, 4 to 2 × 2, 9 to 3 × 3 ; and so on. But, if the reader will consider the matter, he will see that the multiplication of a number (the physical property of an object) by itself does not correspond to any simple relation between the things counted ; by the mere examination of counted objects, we should never be led to consider such an operation at all. It is suggested to us only because we have drawn up our multiplication table and have reached the idea of multiplying one *numeral* by another, irrespective of what is represented by that numeral. We know what is the result of multiplying 3 × 3, when the two 3's represent different numbers and the multiplication corresponds to a physical operation on things counted ; it occurs to us that the multiplication of 3 by *itself*, when the two 3's represent the same thing, although it does not correspond to a physical relation, may yet correspond to the numerical relation in a numerical law. And we find once more that this suggestion turns out to be true ; there are numerical laws in which this numerical relation is found. Thus if we measure (1) the time during which a body starting from rest has been falling (2) the distance through which it has fallen during that time, we should get in our notebook parallel columns like this :

TABLE II

TIME		DISTANCE	TIME		DISTANCE
1	..	1	4	..	16
2	..	4	5	..	25
3	..	9	6	..	36

The numerals in the second column are arrived at by multiplying those in the first by themselves ; in technical language, the second column is the " square " of the first.

Another example. In place of dividing one column by some fixed number in order to get the other, we may use the multiplication table to divide some fixed number (e.g. 1) by that column. Then we should get the table

1	..	1·00	3	..	0·33
2	..	0·50	4	..	0·25
		5	..	0·20	

and so on. Here, again, is a pure numerical operation which does not correspond to any simple physical relation upon numbers ; there is no collection simply related to another collection in such a way that the number of the first is equal to that obtained by dividing 1 by the number of the second. (Indeed, as we have seen that fractions have no application to number, and since this rule must lead to fractions, there cannot be such a relation.) And yet once more we find that this numerical relation does occur in a numerical law. If the first column represented the pressure on a given amount of gas, the second would represent the volume of that gas.

So far, all the relations we have considered were derived directly from the multiplication table. But an extension of the process that we are tracing leads to relations which cannot be derived directly and thus carries us further from the original suggestions indicated by mere counting. Let us return to Table II, and consider what would happen if we found for the numerals in the second column values intermediate between those given. Suppose we measured the distance first and found 2, 3, 5, 6, 7, 8, 10, 11, 12, 13, 14, 15 . . .; what does the rule lead us to expect for the corresponding entries in the first column, the values of the time. The answer will be given if in the multiplication table we can find numerals which, when multiplied by themselves, give 2, 3, 5 . . . But a search will reveal that there are no such numerals. We can find numerals which, when multiplied by themselves give very nearly 2, 3, 5 . . . ; for instance, 1·41, 1·73, 2·24 give 1·9881, 2·9929, 5·0166, and we could find numerals which would come even closer to those desired. And that is really all we want, for our measurements are never perfectly accurate, and if we can get numerals which agree very nearly with our rule, that is all that we can expect. But the search for such numerals would be a very long and tedious business ; it would involve our drawing up an enormously complicated multiplication table, including not only whole numbers but also fractions with many decimal places. And so the question arises if we cannot find some simpler rule for obtaining quickly the number which multiplied by itself will come as close as we please to 2, 3, 4 . . . Well, we can ; the rule is given in every textbook of arithmetic ; it need not be given here. The point which interests us is that, just as the simple multiplication of two numerals suggested a new process, namely the multiplication of a numeral by itself, so this new process suggests in its turn many other and more complicated processes. To each of these new processes corresponds a new rule for relating numerals and for arriving at one starting from another ; and to each new rule may correspond a numerical law. We thus get many fresh forms of numerical law suggested, and some of them will be found to represent actual experiments.

This process for extending arithmetical operations beyond the simple division and multiplication from which we start ; the consequent invention of new rules for relating numerals and deriving one from another ; and the study of the rules, when they are invented—all this is a purely intellectual process. It does not depend on experiment at all ; experiment enters only when we inquire whether there is an actual experimental law stating one of the invented numerical relations between measured properties. The process is, in fact, part of mathematics, not of experimental science ; and one of the reasons why mathematics is useful to science is that it suggests possible new forms for numerical laws. Of course the examples that have been given are extremely elementary, and the actual mathematics of to-day has diverged very widely from such simple considerations ; but the invention of such rules leads, logically if not historically, to one of the great branches of modern mathematics, the Theory of Functions. (When two numbers are related as in our tables, they are technically said to be " functions " of each other.) It has been developed by mathematicians to satisfy their own intellectual needs, their sense of logical neatness and of form ; but though great tracts of it have no bearing whatever upon experimental science, it still remains remarkable how often relations developed by the mathematician for his own purposes prove in the end to have direct and immediate application to the experimental facts of science.

NUMERICAL LAWS AND DERIVED MEASUREMENT

In this discussion there has been overlooked temporarily the feature of numerical laws which, in the previous chapter, we decided gave rise to their importance, namely, that they made possible systems of derived measurement. In the first law, taken as an example (Table I), the rule by which the numerals in the second column were derived from those in the first involved a numeral 7, which was not a member of those columns, but an additional number applicable equally to all members of the columns. This constant numeral, characteristic of the rule asserted by the numerical law, represented a property of the system investigated and permitted a derived measurement of that system. But in Table II, there is no such constant numeral ; the rule for obtaining the second from the first column is simply that the numerals in the first column are to be multiplied by themselves ; no other numeral is involved. But this simplicity is really misleading ; we should not, except by a mere " fluke," ever get such a table as Table II as a result of our measurements. The reason is this. Suppose that, in obtaining Table II, we have measured the time in seconds and the distance fallen in feet ; and that we now propose to write down the result of exactly the same measurements, measuring the time in minutes and the distances in yards. Then the numerals in the first column, representing exactly the same observations, would all be divided by 60 and those in the second would all be divided by 3 ; the observation which was represented before by 60 in the first column would now be represented by 1 ; and the number in the second column represented before by 3 would now be represented by 1. If I now apply the rule to the two columns I shall find it will not work ; the second is *not* the first multiplied by itself. But there will be a new rule, as the reader may see for himself ; it will be that the second column is the same as the first, when the first is (1) multiplied by itself, and (2) the result multiplied by 1,200. And if we measured the time and the distance in some other units (say hours and miles), we should again have to amend our rule, but it would only differ from the former rule in the substitution for 1,200 of some other numeral. If we choose our units in yet a third way, we should get a third rule, and this time the constant numeral might be 1. We should have exactly Table II ; but we should get that table exactly only because we had chosen our units of time and distance in a particular way.

These considerations are quite general. Whatever the

numerical law, the rule involved in it will be changed by changing the unit in which we measure the properties represented by the two columns ; but the change will only consist in the substitution of one constant numeral for another. If we chance to choose the units in some particular way, that constant numeral may turn out to be 1 and so will disappear from sight. But it will always be there. There must be associated with every numerical law, involving a rule for arriving at the numerals in one column from those in the other, some constant numeral which is applicable to all members of the column alike. And this constant may always, as in the case of density, be the measure of some property to which derived measurement is applicable. Every numerical law therefore—this is the conclusion to be enforced—may give rise to a system of derived measurement ; and as a matter of fact all important numerical laws do actually so give rise.

CALCULATION

But though the establishment of system of derived measurement is one use of numerical laws, they have also another use, which is even more important. They permit *calculation*. This is an extremely important conception which deserves our close attention.

Calculation is the process of combining two or more numerical laws in such a way as to produce a third numerical law. The simplest form of it may be illustrated by the following example. We know the following two laws which, in rather different forms, have been quoted before : (1) the weight of a given volume of any substance is proportional to its density ; (2) the density of a gas is proportional to the pressure upon it. From these two laws we can deduce the third law : the weight of a given volume of any gas is proportional to the pressure upon it.

That conclusion seems to follow directly without any need for further experiments. Accordingly we appear to have arrived at a fresh numerical law without adducing any fresh experimental evidence. But is that possible? All our previous inquiry leads us to believe that laws, whether numerical or other, can only be proved by experimental inquiry and that the proof of a new law without new experimental evidence is impossible. How are we to reconcile the two conclusions? When we have answered that question we shall understand what is the importance of calculation for science.

Let us first note that it is possible, without violating the conclusions already reached, to deduce *something* from a numerical law by a process of mere thought without new experiment. For instance, from the law that the density of iron is 7, I can deduce that a portion of it which has a volume 1 will have a weight 7. But this deduction is merely stating in new terms what was asserted by the original law; when I said that the density of iron was 7, I meant (among other things) that a volume 1 had a weight 7; if I had not meant that I should never have asserted the law. The " deduction " is nothing but a translation of the law (or of part of it), into different language, and is of no greater scientific importance than a translation from (say) English into French. One kind of translation, like the other, may have useful results, but it is not the kind of useful result that is obtained from calculation. Pure deduction never achieves anything but this kind of translation; it never leads to anything new. But the calculation taken as an example does lead to something new. Neither when I asserted the first law, nor when I asserted the second did I mean what is asserted by the third; I might have asserted the first without knowing the second and the second without knowing the first (for I might have known what the density of a gas was under different conditions without knowing precisely how it is measured) ; and I might have asserted either of them, without knowing the third. The third law is not merely an expression in different words of something known before ; it is a new addition to knowledge.

But we have added to knowledge only because we have introduced an assertion which was not contained in the two original statements. The deduction depends on the fact that if one thing (A) is proportional to another thing (B) and if B is proportional to a third thing (C), then A is proportional to C. This proposition was not contained in the original statements. But, the reader may reply, it *was* so contained, because it is involved in the very meaning of " proportional " ; when we say that A is proportional to B, we mean to imply the fact which has just been stated. Now that is perfectly true if we are thinking of the mathematical meaning of " proportional," but it is not true if we are thinking of the physical meaning. The proposition which we have really used in making our deduction is this : If weight is proportional (in the mathematical sense) to density, when weight is varied by taking different substances, then it is also proportional to density when weight is varied by compressing more of the same substance into the same volume. That is a statement which experiment alone can prove, and it is because we have in fact assumed that experimental statement that we have been able to " deduce " a new piece of experimental knowledge. It is involved in the original statements only if, when it is said that density is proportional to pressure, it is implied that it has been ascertained by experiment that the law of density is true, and that there is a constant density of a gas, however compressed, given by dividing the weight by the volume.

The conclusion I want to draw is this. When we appear to arrive at new scientific knowledge by mere deduction from previous knowledge, we are always assuming some experimental fact which is not clearly involved in the original statements. What we usually assume is that some law is true in circumstances rather more general than those we have considered hitherto. Of course the assumption may be quite legitimate, for the great value of laws is that they are applicable to circumstances more general than those of the experiments on which they are based ; but we can never be perfectly sure that it is legitimate until we try. Calculation, then, when it appears to add anything to our knowledge, is always slightly precarious ; like theory, it suggests strongly that some law may be true, rather than proves definitely than some law must be true.

So far we have spoken of calculation as if it were merely deduction ; we have not referred to the fact that calculation always involves a special type of deduction, namely mathematical deduction. For there are, of course, forms of deduction which are not mathematical. All argument is based, or should be based, upon the logical processes which are called deduction ; and most of us are prepared to argue, however slight our mathematical attainments. I do not propose to discuss here generally what are the distinctive characteristics of mathematical argument ; for an exposition of that matter the reader should turn to works in which mathematicians expound their own study.[1] I want only to consider why it is that this kind of deduction has such a special significance for science. And, stated briefly, the reason is this. The assumption, mentioned in the last paragraph, which is introduced in the process of deduction, is usually suggested by the form of the deduction and by the ideas naturally associated with it. (Thus, in the example we took, the assumption is suggested by the proposition quoted about proportionality

[1] E.g. "An Introduction to Mathematics," by Prof. Whitehead, in the Home University Library.

which is the idea especially associated by the form of the deduction). The assumptions thus suggested by mathematical deduction are almost invariably found to be actually true. It is this fact which gives to mathematical deduction its special significance for science.

Again an example is necessary and we will take one which brings us close to the actual use of mathematics in science. Let us return to Table II which gives the relation between the time for which a body has fallen and the distance through which it has fallen. The falling body, like all moving bodies, has a " velocity." By the velocity of a body we mean the distance that it moves in a given time, and we measure the velocity by dividing that distance by that time (as we measure density by dividing the weight by the volume). But this way of measuring velocity gives a definite result only when the velocity is constant, that is to say, when the distance travelled is proportional to the time and the distance travelled in any given time is always the same (compare what was said about density on p. 130). This condition is not fulfilled in our example ; the distance fallen in the first second is 1, in the next 3, in the third 5, in the next 7 —and so on. We usually express that fact by saying that the velocity increases as the body falls ; but we ought really to ask ourselves whether there is such a thing as velocity in this case and whether, therefore, the statement can mean anything. For what is the velocity of the body at the end of the 3rd second—i.e. at the time called 3. We might say that it is to be found by taking the distance travelled in the second before 3, which is 5, or in the second after 3, which is 7, or in the second of which the instant " 3 " is the middle (from $2\frac{1}{2}$ to $3\frac{1}{2}$), which turns out to be 6. Or again we might say it is to be found by taking *half* the distance travelled in the two seconds of which " 3 " is the middle (from 2 to 4) which is again 6. We get different values for the velocity according to which of these alternatives we adopt. There are doubtless good reasons in this example for choosing the alternative 6, for two ways (and really many more than two ways, all of them plausible) lead to the same result. But if we took a more complicated relation between time and distance than that of Table II, we should find that these two ways gave different results, and that neither of them were obviously more plausible than any alternative. Do then we mean anything by velocity in such cases and, if so, what do we mean ?

It is here that mathematics can help us. By simply thinking about the matter Newton, the greatest of mathematicians, devised a rule by which he suggested that velocity might be measured in all such cases.[1] It is a rule applicable to every kind of relation between time and distance that actually occurs ; and it gives the " plausible " result whenever that relation is so simple that one rule is more plausible than another. Moreover it is a very pretty and ingenious rule ; it is based on ideas which are themselves attractive and in every way it appeals to the æsthetic sense of the mathematician. It enables us, when we know the relation between time and distance, to measure uniquely and certainly the velocity at every instant, in however complicated a way the velocity may be changing. It is therefore strongly suggested that we take as the velocity the value obtained according to this rule.

But can there be any question whether we are right or wrong to take that value ; can experiment show that we ought to take one value rather than another ? Yes, it

[1] This is part of the great mathematical achievement mentioned on p. 100. I purposely refrain from giving the rule, not because it is really hard to explain, but because I want to make clear that what is important is to have *some* rule, not any particular rule.

can; and in this way. When the velocity is constant
and we can measure it without ambiguity, then we can
establish laws between that velocity and certain properties
of the moving body. Thus, if we allow a moving steel ball
to impinge on a lead block, it will make a dent in it determined by its velocity ; and when we have established by
observations of this kind a relation between the velocity
and the size of the dent, we can obviously use the size of
the dent to measure the velocity. Suppose now our
falling body is a steel ball, and we allow it to impinge
on a lead block after falling through different distances ;
we shall find that its velocity, estimated by the size of
the dent, agrees exactly with the velocity estimated by
Newton's rule, and not with that estimated by any other
rule (so long, of course, as the other rule does not give the
same result as Newton's). That, I hope the reader will
agree, is a very definite proof that Newton's rule is right.

On this account only Newton's rule would be very
important, but it has a wider and much more important
application. So far we have expressed the rule as giving
the velocity at any instant when the relation between
time and distance is known ; but the problem might be
reversed. We might know the velocity at any instant
and want to find out how far the body has moved in any
given time. If the velocity were the same at all instants,
the problem would be easy ; the distance would be the
velocity multiplied by the time. But if it is not the same,
the right answer is by no means easy to obtain ; in fact
the only way of obtaining it is by the use of Newton's
rule. The form of that rule makes it easy to reverse it
and, instead of obtaining the velocity from the distance,
to obtain the distance from the velocity ; but until that
rule was given, the problem could not have been solved ;
it would have baffled the wisest philosophers of Greece.
Now this particular problem is not of any very great
importance, for it would be easier to measure by experi-

ment the distance moved than to measure the velocity and calculate the distance. But there are closely analogous cases—one of which we shall notice immediately—in which the position will be reversed. Let us therefore ask what is the assumption which, in accordance with the conclusion reached on p. 146, must be introduced, if the solution of the problem is to give new experimental knowledge.

We have seen that the problem could be solved easily if the velocity were constant ; what we are asking, is how it is to be solved if the velocity does not remain constant. If we examined the rule by which the solution is obtained, we should find that it involves the assumption that the effect upon the distance travelled of a certain velocity at a given instant of time is the same as it would be if the body had at that instant the same *constant* velocity. We know how far the body would travel at that instant if the velocity were constant, and the assumption tells us that it will travel at that instant the same distance although the velocity is not constant. To obtain the whole distance travelled in any given time, we have to add up the distances travelled at the instants of which that time is made up ; the reversed Newtonian rule gives a simple and direct method for adding up these distances, and thus solves the problem. It should be noted that the assumption is one that cannot possibly be proved by experiment ; we are assuming that something would happen if things were not what they actually are ; and experiment can only tell us about things as they are. Accordingly calculation of this kind must, in all strictness, always be confirmed by experiment before it is certain. But as a matter of fact, the assumption is one of which we are almost more certain than we are of any experiment. It is characteristic, not only of the particular example that we have been considering, but of the whole structure of modern mathematical physics which has arisen out of the work of Newton. We should never think it really necessary to-day to confirm by experiment the results of calculation based on that assumption ; indeed if experiment and calculation did not agree, we should always maintain that the former and not the latter was wrong. But the assumption is there, and it is primarily suggested by the æsthetic sense of the mathematician, not dictated by the facts of the external world. Its certainty is yet one more striking instance of the conformity of the external world with our desires.

And now let us glance at an example in which such calculation becomes of real importance. Let us take a pendulum, consisting of a heavy bob at the end of a pivoted rod, draw it aside and then let it swing. We ask how it will swing, what positions the bob will occupy at various times after it is started. Our calculation proceeds from two known laws. (1) We know how the force on the pendulum varies with its position. That we can find out by actual experiments. We hang a weight by a string over a pulley, attach the other end of the string to the bob, and notice how far the bob is pulled aside by various weights hanging at the end of the string. We thus get a numerical law between the force and the angle which the rod of the pendulum makes with the vertical. (2) We know how a body will move under a constant force. It will move in accordance with Table II, the distance travelled being proportional to the " square " of the time during which the force acts. Now we introduce the Newtonian assumption. We know the force in each position ; we know how it would move in that position if the force on it were constant ; actually it is not constant, but we assume that the motion will be the same as it would be if, in that position, the force were constant. With that assumption, the general Newtonian rule (of which the application to velocity is only a special instance) enables us to sum up the effects of the motions in the different positions, and thus to arrive at the desired relation between the time and the positions successively occupied by the pendulum. The whole of the calculation which plays so large a part in modern science is nothing but an elaboration of that simple example.

MATHEMATICAL THEORIES

We have now examined two of the applications of mathematics to science. Both of them depend on the fact that relations which appeal to the sense of the mathematician by their neatness and simplicity are found to be important in the external world of experiment. The relations between numerals which he suggests are found to occur in numerical laws, and the assumptions which are suggested by his arguments are found to be true. We have finally to notice a yet more striking example of the same fact, and one which is much more difficult to explain to the layman.

This last application is in formulating theories. In Chapter V we concluded that a theory, to be valuable, must have two features. It must be such that laws can be predicted from it and such that it explains these laws by introducing some analogy based on laws more familiar than those to be explained. In recent developments of physics, theories have been developed which conform to the first of these conditions but not to the second. In place of the analogy with familiar laws, there appears the new principle of mathematical simplicity. These theories explain the laws, as do the older theories, by replacing less acceptable by more acceptable ideas; but the greater acceptability of the ideas introduced by the theories is not derived from an analogy with familiar laws, but simply from the strong appeal they make to the mathematician's sense of form.

I do not feel confident that I can explain the matter further to those who have not some knowledge of both physics and mathematics, but I must try.   The laws on the analogy with which theories of the older type are based were often (in physics, usually) numerical laws, such laws for example as that of the falling body.   Now numerical laws, since they involve mathematical relations, are usually expressed, not in words, but in the symbols in which, as every one knows, mathematicians express their ideas and their arguments.   I have been careful to avoid these symbols ; until this page there is hardly an " x " or a " y " in the book.   And I have done so because experience shows that they frighten people ; they make them think that something very difficult is involved. But really, of course, symbols make things easier ; it is conceivable that some super-human intellect might be able to study mathematics, and even to advance it, expressing all his thoughts in words.   Actually, the wonderful symbolism mathematics has invented make such efforts unnecessary ; they make the processes of reasoning quite easy to follow.   They are actually inseparable from mathematics ; they make exceedingly difficult arguments easy to follow by means of simple rules for juggling with these symbols—interchanging their order, replacing one by another, and so on.   The consequence is that the expert mathematician has a sense about symbols, as symbols ; he looks at a page covered with what, to anyone else, are unintelligible scrawls of ink, and he immediately realizes whether the argument expressed by them is such as is likely to satisfy his sense of form ;   whether the argument will be " neat " and the results " pretty."   (I can't tell you what those terms mean, any more than I can tell you what I mean when I say that a picture is beautiful.)

Now sometimes, but not always, simple folk can understand what he means ; let me try an example.

Suppose you found a page with the following marks on it—never mind if they mean anything :

$$i = \frac{d\gamma}{dy} - \frac{d\beta}{dz} \qquad \frac{dX}{dt} = \frac{d\gamma}{dy} - \frac{d\beta}{dz}$$

$$j = \frac{d\alpha}{dz} - \frac{d\gamma}{dx} \qquad \frac{dY}{dt} = \frac{d\alpha}{dz} - \frac{d\gamma}{dx}$$

$$k = \frac{d\beta}{dx} - \frac{d\alpha}{dy} \qquad \frac{dZ}{dt} = \frac{d\beta}{dx} - \frac{d\alpha}{dy}$$

$$\frac{d\alpha}{dt} = \frac{dY}{dz} - \frac{dZ}{dy} \qquad \frac{d\alpha}{dt} = \frac{dY}{dz} - \frac{dZ}{dy}$$

$$\frac{d\beta}{dt} = \frac{dZ}{dx} - \frac{dX}{dz} \qquad \frac{d\beta}{dt} = \frac{dZ}{dx} - \frac{dX}{dz}$$

$$\frac{d\gamma}{dt} = \frac{dX}{dy} - \frac{dY}{dx} \qquad \frac{d\gamma}{dt} = \frac{dX}{dy} - \frac{dY}{dx}$$

I think you would see that the set of symbols on the right side are " prettier " in some sense than those on the left ; they are more symmetrical. Well, the great physicist, James Clerk Maxwell, about 1870, thought so too ; and by substituting the symbols on the right side for those on the left, he founded modern physics, and, among other practical results, made wireless telegraphy possible.

It sounds incredible ; and I must try to explain a little more. The symbols on the left side represent two well-known electrical laws : Ampère's Law and Faraday's Law ; or rather a theory suggested by an analogy with those laws. The symbols i, j, k represent in those laws an electric current. For these symbols Maxwell substituted $\dfrac{dX}{dt} \ \dfrac{dY}{dt} \ \dfrac{dZ}{dt}$; that substitution was roughly equivalent to saying that an electric current was related to the things represented by X, Y, Z, t (never mind what they are) in a way nobody had ever thought of before ; it was equivalent to saying that so long as X, Y, Z, t were related in a certain way, there might be an electric current in circumstances in which nobody had believed that an electric current could flow. As a matter of fact, such a current would be one flowing in an absolutely empty space without any material conductor along which it might flow, and such a current was previously thought to be impossible. But Maxwell's feeling for symbolism suggested to him that there might be such a current, and when he worked out the consequences of supposing that there were such currents (not currents perceptible in the ordinary way, but theoretical currents, as molecules are theoretical hard particles), he arrived at the unexpected result that an alteration in an electric current in one place would be reproduced at another far distant from it by waves travelling from one to the other through absolutely empty space between. Hertz actually produced and detected such waves ; and Marconi made them a commercial article.

That is the best attempt I can make at explaining the matter. It is one more illustration of the marvellous power of pure thought, aiming only at the satisfaction of intellectual desires, to control the external world. Since Maxwell's time, there have been many equally wonderful theories, the form of which is suggested by nothing but the mathematician's sense for symbols. The latest are those of Sommerfeld, based the ideas of Niels Bohr, and of Einstein. Every one has heard of the latter, but the former (which concerns the constitution of the atom) is quite as marvellous. But of these I could not give, even if space allowed, even such an explanation as I have attempted for Maxwell's. And the reason is this : A theory by itself means nothing experimental—we insisted on that in Chapter V—it is only when something is deduced from it that it is brought within the range of our material senses. Now in Maxwell's theory, the symbols, in the alteration of which the characteristic feature of the theory depends, are retained through the deduction and appear in the law which is compared with experiment. Accordingly it is possible to give some idea of what these symbols mean in terms of things experimentally observed. But in Sommerfeld's or Einstein's theory the symbols, which are necessarily involved in the assumption which differentiates their theories from others, disappear during the deduction ; they leave a mark on the other symbols which remain and alter the relation between them ; but the symbols on the relations of which the whole theory hangs, do not appear at all in any law deduced from the theory. It is quite impossible to give any idea of what they mean in terms of experiment.[1] Probably some of my readers will have read the very interesting and ingenious attempts to " explain Einstein " which have been published, and will feel that they really have a grasp of the matter. Personally I doubt it ; the only way to understand what Einstein did is to look at the symbols in which his theory must ultimately be expressed and to realize that it was reasons of symbolic form, and such reasons alone, which led him to arrange the symbols in the way he did and in no other.

But now I have waded into such deep water that it is time to retrace my steps and return to the safe shore of the affairs of practical life.

[1] The same is true really of the exposition of the Newtonian assumption attempted on p. 151. It is strictly impossible to state exactly what is the assumption discussed there without using symbols. The acute reader will have guessed already that on that page I felt myself skating on very thin ice.

CHAPTER VIII

THE APPLICATIONS OF SCIENCE

THE PRACTICAL VALUE OF SCIENCE

SO far we have regarded science as a means of satis-fying our purely intellectual desires. And it must be insisted once again that such is the primary and fundamental object of science; if science did not fulfil that purpose, then it could certainly fulfil no other. It has applications to practical life, only because it is true; and its truth arises directly and immediately from its success as an instrument of intellectual satis-faction. Nevertheless there is no doubt that, for the world at large—the world which includes those to which this book is addressed—it is the practical rather than the intellectual value of science which makes the greater appeal. I do not mean that they are blind to the things of the mind, and consider only those of the body; I mean merely that science is not for them the most suitable instrument by which they may cultivate their minds. Art, history, and philosophy are competing vehicles of culture; and their sense of the supreme value of their own study should not lead men of science to insist that its value is unique. Indeed, if we are forced to recognize that pure science will always be an esoteric study, it should increase our pride that we are to be found in the inner circle of the elect. On the other hand, since man cannot live by thought alone, the practical value of science makes a universal appeal; it would be pedantic and misleading to omit some consideration of this aspect of science.

The practical value of science arises, of course, from the formulation of laws. Laws predict the behaviour of that external world with which our practical and everyday life is an unceasing struggle. Forewarned is forearmed, and we stand a better chance of success in the contest if we know precisely how our adversary may be expected to behave. Knowledge is power and our knowledge of the external world enables us in some measure to control it. So much is obvious ; nobody to-day will be found to deny that science—and it must be remembered that we always use that word to denote the abstract study—might be of great service in practical life. Nor indeed will anybody deny that it has been of great service. We have all heard how the invention of the dynamo—on which is based every industrial use of electricity, without which modern civilization would be impossible—or the discovery of the true nature of ferments—the basis of modern medicine—was the direct outcome of the purest and most disinterested intellectual inquiries. But although this is granted universally, men of science are still heard to complain with ever-increasing vehemence that they are not allotted their due share of influence in the control of industry and of the State, and that science is always suffering from material starvation. It is clear, therefore, that in spite of the superficial agreement on the value of science, there is still an underlying difference of opinion which merits our attention.

The difference is not surprising, for candour compels us to confess that these admitted facts, on which the claims of science to practical value are often based, are not really an adequate basis for those claims. The fact that science might produce valuable results and actually has produced some, is no more justification for our devoting any great part of our energies to its development than the fact that I picked up half a crown in the street yesterday—and might pick up another—would be a justification for my abandoning sober work to search for buried treasure. Moreover, the very people who claim, on the ground of the work of Faraday or Pasteur, that science should receive large endowments and a great share in government, often urge at the same time that Faraday and Pasteur were examples of that genius which cannot be produced by training and can scarcely be stunted by adversity. If it were only these exceptional achievements, occurring two or three times a century, which had practical value, the encouragement of science would be an unprofitable gamble. If we are really to convince the outside world of the need for the closer application of science to practical affairs, we must give reasons for our claim much more carefully and guardedly than has been the custom up to the present. Nothing is more fatal to our cause than to encourage expectations doomed to disappointment.

Accordingly in this chapter, I propose to diverge entirely from the usual path. I shall not give a single example of practical science. There are plenty of good books which tell what science has achieved in the past, and plenty of newspaper paragraphs to tell us what it is going to achieve in the future. Here I want to inquire carefully what value science might have for practical life, why it has that value, and under what conditions its value is most likely to be realized.

THE LIMITATIONS OF SCIENCE

It will be well to point out immediately that science, like everything else, has its limitations. and that there are some practical problems which, from its very nature, it is debarred from solving. It must never be forgotten that, though science helps us in controlling the external world, it does not give us the smallest indication in what direction that control should be exercised. Whenever we undertake any practical action, we have two decisions to make ; we have to decide what is the *end* of our action, what result we wish to obtain ; and we have to decide what is the right *means* to that end, what action will produce the desired result.   The distinction between the two decisions can be traced in the simplest as well as (perhaps better than) in the most complex actions.   If I go to a meal in a restaurant, I have first to decide whether I want beef or mutton, tea or coffee, and second how I am to get what I want.   If I have toothache, I have first to decide whether I want to be cured, and second if I am more likely to be cured if I doctor myself or if I go to a dentist.   The fact that there are two decisions is sometimes obscured by the simplicity and obviousness of one of them.   In the first example, the decision as to means is liable to be overlooked ; for (except in some restaurants) it is obvious that the best way to get the meal I want is to ask for it.   In the second, the decision as to end may be unobserved, because it is so obvious that I want to be cured.

In these simple examples the distinction between the two decisions is clear ; in others they are so closely interconnected that care is needed to separate them.   Our choice of the ends at which we may aim is often determined in part by the means we have of attaining them ; it is foolish to struggle towards a goal that can never be reached.   On the other hand, action which is desirable as a means to one end, may be objectionable because it leads at the same time to other results that are undesirable as ends.   In all the more complicated decisions of life, such conflicts between ends and means arise, and it is a necessary step towards accuracy of thought to disentangle the conflicting elements.   It is all the more necessary, because in controversial matters there is always a tendency to conceal questions of ends and to pretend that every question is one of means only ; the reason is that agreement concerning ends is far less easily attainable than agreement concerning means, so that, when we are trying to make converts to our views, we are naturally apt to disguise differences that are irreconcilable.

Political discussion provides examples of this tendency. It is clap-trap to announce portentously that we all desire the welfare of the community and to pretend that we differ only in our view of the best way of attaining it ; what we really differ about is our ideas of the welfare of the community ; we disagree as to what is the state of society that forms the end of our political action. If we could agree about that, our remaining differences would not excite much heat. As it is, our pretence that we are arguing merely about means often leads us to adopt means which are obviously ill-adapted to secure any of the ends at which any of the contending parties aim.

Since science must always exclude from its province judgments concerning which differences are irreconcilable, it can only guide practical life in the choice of means, and not in the choice of ends. If one course of action is more " scientific " than another, that course is better only in the sense that is a more efficient means to some end ; from the fact that it is indicated as a result of scientific inquiry, it is quite illegitimate to conclude that the action must necessarily be desirable. That conclusion follows only if it can be proved that the end, to which the action is a means, is desirable ; such a proof must always lie wholly without the range of science. The neglect to observe this distinction is responsible for much of the disregard, and even actual dislike, of their study in its application to practical life of which men of science so often complain. It was seriously urged in recent years that science, being responsible for the horrors of modern warfare, is a danger to civilization ; and I am told that many manual workers are inclined to regard science with hostility because it is associated in their minds with the

" scientific " management of industry.[1]   Such objections
are altogether unjust ; science gives to mankind a greater
power of control over his environment.   He may use that
control for good or for ill ; and if he uses it for ill, the fault
lies in that part of human nature which is most remote
from science ; it lies in the free exercise of will.   To deny
knowledge for fear it may be misused is to repeat the
errors of the mediæval church ; thus to deprive men of
the power to do evil is to deprive them of the yet greater
power to do good.   For precisely the same knowledge
that has made Europe a desert has given the power to
restore her former fertility ; and the increase of individual
productive power which may be used to rivet more closely
the chains of wage-slavery might also give to the worker
that leisure from material production which alone can
give freedom to the slave.

Men of science themselves are largely to blame for the
confusion against which this protest is directed.   They
are so accustomed to having to force their conclusions on
an ignorant and reluctant world, that they are apt to
overstep the limits of their special sphere ; they sometimes forget that they cease to be experts when they
leave their laboratories, and that in deciding questions
foreign to science, they have no more (but, of course,
no less) claim to attention than anyone else.   Like the
members of any other trade or profession, they are apt
to be affected in their social and political views by the
work which is their main occupation, and to lay special
stress on the evils which come immediately under their
notice.[2]   In this respect it is useless to expect them to

[1] We need not discuss here whether the methods of factory control
to which objection is taken have really any claim to be scientific in our
sense ; whether, that is to say, they are the outcome of such investigation as has been described in the earlier chapters.

[2] I am tempted to describe what are the social and political views
which the study and practice of science tends to inculcate.   But this
is a matter in which the spectator sees most of the game ; if I made
the attempt, I should probably be led astray by my own particular
opinions.

be more perfect than the rest of mankind. But any danger of paying too much or too little heed to pronouncements put forward on behalf of " science " will be avoided if the distinction on which so much stress has been laid is borne in mind. On questions of means to a given end (if they concern the nature of the external world) science is the one and only true guide ; on questions of the ends to which means should be directed, science has nothing to say.

THE CERTAINTY OF SCIENTIFIC KNOWLEDGE

I have thought it better thus to start with a consideration of the limitations of science ; not because the greater danger lies in the neglect of those limitations, but merely to convince the reader that I am not blind to their existence. Actually, in this country at least, the greater danger lies in the other direction, in refusing to accept the clear and positive decisions of science on matters which lie wholly within its bounds. Why is there any such danger ? It arises, I believe, from two sources not wholly independent. The first source is a disbelief that science is really possessed of any definite knowledge. Scientific experts seem to differ as much as experts in other subjects, and may be heard in any patent litigation swearing cheerfully against each other. The second source is a general distrust of the " theorist " as compared with the " practical man." The chief points that have to be raised will appear naturally in a discussion of these two errors.

It may be thought that the first " error " has been implicitly confirmed by our previous discussions. For it has been urged that there is a strong personal element in science and that complete agreement is to be found only in its subject matter and not in its conclusions. But while it is perfectly true that a theory, and even to some extent a law, may be an object of contention when it is first proposed, it is equally true that the difference of opinion is always ultimately resolved.   A theory may be doubted, but while it is doubted it is not part of the firm fabric of science ; but in the end it is always either definitely accepted or definitely rejected.   It is in this that science differs from such studies as history or philosophy in which controversies are perennial.   There is an immense body of science concerning which there is no doubt, and that body includes both theories and laws ; there is a smaller part concerning which dispute is still continuing.   It is only natural that this smaller part should receive the greater share of explicit attention ; the other and greater part we learn in our school and university courses and find no need to discuss later, because it is a matter of common knowledge with which all properly informed persons are completely familiar ; it is the base from which we proceed to establish new knowledge, and the premiss on which we found our arguments concerning it.   The distinction between the two parts of scientific knowledge, that which is firmly established, and that which is still doubtful, is perfectly clear and definite to all who have been properly trained.   The fact that doctors differ in science, as in other things, does not affect the equally important fact that in much the larger part of their knowledge they agree.

But a more serious objection may be raised.   In the opening chapters we concluded that science draws its subject-matter from a limited portion of experience, and that this limited portion necessarily excludes all that part of our life which is of the most intimate interest to us. It may be urged with force that while science may be in possession of perfectly positive knowledge concerning which every one who has studied the matter is in agreement, yet this knowledge is entirely divorced from all the affairs of practical life ; when science attempts to intrude into such affairs it becomes as hesitating and dubitable as any other source of knowledge.

As a formal statement of the position, this objection must be admitted as valid. The uniformly certain and completely universal laws of science can be realized only in the carefully guarded conditions of the laboratory, and are never found in the busy world outside. There is scarcely any event or process of practical importance to which we could point as providing a direct confirmation of any of the propositions of pure science, or which could be described completely in terms of those propositions. In every such event and process, there is involved some element of which science can take no cognizance, and it is usually on account of this element (as was remarked in Chapter III) that the event has practical importance. And again, it is the presence of this element which makes it possible for experts, equally well-informed, to differ in their preliminary suggestions of an explanation of the event or of the most suitable means for controlling it. But it does not follow because practical events do not lie wholly within the realm of science that they lie wholly without it. Indeed it is from the study of practically important events that many of the results of pure science have actually been derived. Let us examine the matter more carefully.

All the applications of science to practical life depend ultimately on a knowledge of laws. Whether we are asked to explain an event, or to suggest means whereby an event may be produced or prevented, we can meet the demand only if we know the laws of which the event is the consequence. But laws state only relations between events ; when we say that an event is the consequence of certain laws, we do not mean that this event must happen in all possible circumstances ; we mean only that it is invariably associated with certain other events, and must happen if they happen. The event in question is not only a consequence of the laws, it is a consequence of the laws and of the other events to which it is related by the laws.   Again, it must be noted that I have spoken of laws, not " a law."   The practical event to which our attention is directed will not be a simple event such as is related by pure science to another simple event ; it will be an immensely complicated collection of such events, and these constituent events will be each related to some other event by a separate law.   The constituent events will not in general be related to each other by a law, nor will the other events, to which they are so related, be related to each other by a law.   The explanation of the events in question will not be complete when we have stated a single law, or even all the many laws in which the constituent events are involved ; it is necessary to add that the many events to which they are related by these many laws have actually occurred, and that the many laws are actually in operation.

The last part of the explanation is the part which is not pure science.   Science when it asserts laws, only asserts that, if so-and-so happens, something else will also happen.   But in practical matters it is necessary to convert this hypothetical statement into a definite statement, and to assert that something actually has happened. This is often an extremely difficult matter, which may be the subject of much difference of opinion until all the circumstances have been investigated.   An obvious example of this difficulty arises in the practice of medicine ; diagnosis, the determination of what is wrong with the patient, is a necessary preliminary to his treatment, and is actually the gravest problem with which the physician is faced.   And similar problems arise in all other branches of applied science.   If we are asked to produce some desired product or to find out why the product of some existing process is not satisfactory, the first part of our task must always be to find out exactly in what the desired product consists, and exactly in what particulars it differs from the unsatisfactory product. This problem of determining precisely what are the existing facts is not strictly one of science at all ; the solution of it does not involve the statement of any scientific laws, for laws assert, not what does actually occur, but what will occur if something else occurs. Nevertheless science and scientific laws are useful, and even indispensable, in the solving of it ; for very often the best or only proof that something has occurred is that some other event has occurred with which the first is associated by a law. Thus, the physician bases his diagnosis on his examination of symptoms ; he observes that the bodily state of his patient is abnormal in some particulars, and from his knowledge of the laws connecting those parts of the body which are accessible to observation with those that are not deduces what must be the state of the hidden organs. In the same way the works chemist or physicist is often led to judge what is the source of failure in a product, by examining carefully the process by which it has been produced, and deducing by his knowledge of the laws of chemistry or physics what must be the result of that process.

It is for such reasons that pure science, although it takes no direct cognizance of the actual events of practical life, is of inestimable service in explaining and controlling them. Even though it is impossible to analyse those events completely into laws, it is only by carrying that analysis as far as it will go, and by bringing to light all the laws that are involved, that any explanation or any control can be attained.

These considerations have been suggested by the first of the two errors noted on page 164 ; they answer the objection that science is not possessed of any positive and certain knowledge, or that, if it has such knowledge, it is not relevant to practical problems. The second error is even more dangerous. Science, it is often urged (perhaps not in these actual words), is all very well ; it may even be indispensable ; but it must be the right kind of science. The kind of science that is needed in everyday life is not that of the pure theorist, but that which every practical man is bound to acquire for himself in the ordinary conduct of his business.

Again, it will be well to begin by admitting that there is some truth in the contention that the practical man is likely to manage the business in which he has been immersed all his life better than one who has no experience of any conditions more complicated than those of the laboratory. No doubt scientific men of great eminence often prove as great failures in industry as commercial men would in pure science. But we have already noticed that no practical problem is wholly scientific ; there are questions of ends as well as of means. The scientific man in industry is doubtless apt to be led astray by forgetting that the object of industry is to produce goods, and that processes, however scientifically interesting they may be, are commercially worthless unless they decrease the expenditure of capital and labour necessary to obtain a given amount of goods. Again, at the present time at least, all questions of means have not been brought within the range of science ; the estimation of demand and the foreseeing of supply are matters not yet reduced to any scientific basis. Besides, no man is expert in all sciences, and the fact that he is familiar with one may tend to hide from him his ignorance of another. All this may be readily granted ; but all it proves is that something besides scientific knowledge is required for the competent conduct of affairs. Because the man of science needs the help of the man trained in commerce or administration, it does not follow that the latter does not need the help of the former.

The attack on the practical value of science that we

are considering is best met by a counter-offensive. It sounds plausible to maintain that those who have had the greatest experience of any matter must know most about it. But, like many other plausible doctrines, this one is absolutely false. No popular saying is more misleading than that we learn from experience ; really the capacity of learning from experience is one of the rarest gifts of genius, attained by humble folk only by long and arduous training. Anyone who examines carefully any subject concerning which popular beliefs are prevalent, will always discover that those beliefs are almost uniformly contradicted by the commonest everyday experience. We shall not waste our time if we devote a few pages to discovering the sources of popular fallacies, and considering in what manner they can be corrected by scientific investigation. When we have established how little worthy of confidence is " practical knowledge " we shall be in a position to see the value of " theory."

POPULAR FALLACIES

The most frequent source of such fallacies is a disposition to accept without inquiry statements made by other people. Error from this source is not wholly avoidable ; except in the very few matters in which we can interest ourselves, we must, if we are to avoid blank ignorance, simply believe what we are told by the best authority we can secure. And since nobody is always right, we shall always believe some false doctrines however carefully we choose our authority. But it is very remarkable how people will go on believing things on authority, when the weight of that authority is quite unknown, and when their belief is flatly contradicted by experience. I know a family, not without intelligence, who, until their statement was challenged in a heated argument, always believed, on the authority of some family tradition of unknown origin, that the walk they took every Sunday afternoon was eight miles long ; and yet a party which was not specially athletic, starting after three, always accomplished it before five. A glance at a map which was hanging in their house would have shown them it was barely six miles. This will doubtless seem a very extreme example concerning a very trivial matter, but parallels can be found readily in matters of considerable importance. During the war it was almost a sufficient reason for the army chiefs to adopt some device that it was known (or more often believed) that the enemy made use of it ; and anyone who comes into contact with unscientific managers of industry will be amazed to find how largely their practice is based on hearsay information, and how little evidence they have that the information was reliable or even given in good faith.

In matters which lie outside their own sphere, men of science are often as credulous as anyone else ; but in that sphere, if they are really men of science, intimately acquainted with their study by the actual practice of it, they cannot fail to have learnt how dangerous it is to believe any statement, however firmly asserted by high authority, unless they have tested it for themselves. The necessity for the obtaining of information by direct experiment is embedded in their nature, and no information attained by other means will satisfy them permanently. The determination to believe what is true, and not what other people assert to be true, is the first and not the least important correction applied by science to popular errors.

But if there were no other source of error, reliance on hearsay would not be so dangerous, for our informants would not be so likely to be wrong. There would still be the possibility that they intended to mislead us, though we may neglect that possibility for our purpose. A more serious possibility would remain, namely, that we had misinterpreted their information, and this is actually the greatest danger in knowledge acquired at second-hand. Thus the error about the length of the walk quoted just now, doubtless arose from the fact that the original Sunday walk was eight miles, and that a weaker generation had abbreviated it to six. However, there are other sources of error ; people do draw false conclusions directly from experience ; and even if we could be sure that we had rightly understood an honest informant, there would still be a danger that his information was wrong. In discussing these other sources and giving examples of them, it will be impossible to distinguish them wholly from the first, for all popular beliefs (from which many technical and professional beliefs do not differ essentially) derive much of their weight from their general prevalence. We can only ask what fallacies predisposed men to these beliefs, and thus enabled the beliefs to become prevalent.

The most prolific of these fallacies are false theories. In discussing scientific theories in Chapter V, we saw that in their ultimate nature they are not very different from any other and unscientific attempt to explain things. A theory which suggests that A and B *might* be connected predisposes to the belief that A and B *are* connected. An extreme example of unscientific theories is provided by the superstitions and magical beliefs of primitive civilization. They have ceased to be held explicitly, but they still exert an influence, which is not generally appreciated, on popular beliefs. Thus many people believe, and are very indignant when the assertion is denied, that the poker laid across the bars of the grate will draw up the fire. The belief is based on the old doctrine of the magical power of the cross, formed in this instance by the poker and the bars ; old people can still be found who will say that the poker " keeps the witch up the chimney." Experiment would show that the poker has no effect whatsoever ; but it is not easy to undertake seriously, because the circumstances of " drawing up " a fire are so indeterminate. Most popular weather-lore has a similar origin in false theory ; people are ready to think that the weather will change when the moon changes, only because they think that the moon *might* have some effect on the weather. Again, the persistent feeling that there is some intimate connexion between names, and the things of which they are names, leads to curious credulity. " Rain before seven, fine before eleven," would never have become a popular saying had it not been for the purely verbal jingle.[1]

It is scarcely too much to say that every popular belief concerning such matters is false, and can be refuted by experience which is directly accessible to those who assert it ; and that the reason why such beliefs have gained a hold can always be traced to some false theory which, nowadays, only needs to be expressed to be rejected. Their prevalence indicates clearly how little the majority of mankind can be trusted to analyse their experience carefully, and to base conclusions on that experience and on nothing else. And it has been admitted that the most careful and accurate science does not base its conclusions only on experience ; in science, too, analysis is guided by theory. But there is this vast difference : though science may in the first instance analyse experience and put forward laws guided by theory and not by simple examination of facts, when the analysis is completed and the laws suggested, it does return and compare them with the facts. It is by this

---

[1] There may be this truth in the saying that, in many parts of England, continuous rain for four hours is unusual at any time of the day or night. But if it is meant, as it usually is meant, that rain at seven is more likely to be followed by fine weather than rain at six or eight, then the saying is certainly false.

procedure that it has been able to establish true theories which may be trusted, provisionally at least, in new analysis. The practical man is apt to sneer at the theorist ; but an examination of any of his most firmly-rooted prejudices would show at once that he himself is as much a theorist as the purest and most academic student ; theory is a necessary instrument of thought in disentangling the amazingly complex relations of the external world. But while his theories are false because he never tests them properly, the theories of science are continually under constant test and only survive if they are true. It is the practical man and not the student of pure science who is guilty of relying on extravagant speculation, unchecked by comparison with solid fact.

Closely connected with the errors of false theories are those which arise from false, or more often incomplete, laws. Such laws are, of course, in themselves errors, but they often breed errors much more serious. For we have seen that the things between which laws assert relations are themselves interconnected by laws ; if we start with false laws we are sure to interpret our experience on the wrong lines, because the things between which we shall try to find laws will be such that no laws can involve them. An example which we have used before will make this source of error clear. The word " steel " is used in all but the strictest scientific circles to denote many different things ; or, in other words, there is no law asserting the association of all the properties of all the things which are conventionally called steel. Accordingly there can be no law, strictly true, asserting anything about steel ; for though a law may be found which is true of many kinds of steel, some kind of " steel " can almost certainly be found of which it is not true. If we want to find true laws involving the materials all of which are conventionally termed steel, we must first differentiate the various kinds of steel and seek laws which involve each kind separately.

Neglect of this precaution is one of the most frequent causes of a failure to detect and to cure troubles encountered in industrial processes. The unscientific manager regards as identical everything that is sold to him as steel ; he regards as " water " .everything that comes out of the water main ; and as " gas " everything that comes out of the gas main. He does not realize that these substances, though called by the same name, may have very different properties ; and when his customary process does not lead to the usual result, he will probably waste a great deal of time and money on far-fetched ideas before he realizes that he did not get the same result because he did not start with the same materials. He can expect his processes to be governed by laws and to lead invariably to the same result only if all the materials and operations involved in those laws and employed in those processes are themselves invariable ; that is to say, if their constituent properties and events are themselves associated by invariable laws. This is a very obvious conclusion when it is pointed out, but there is no conclusion more difficult to impress finally on the practical man ignorant of science. He is misled by words. Words are very useful when they really represent ideas, but are a most terrible danger when they do not. A word represents ideas, in the sense important for the practical applications of science, only when the things which it is used to denote are truly collections of properties or events associated by laws ; for it is only then that the word can properly occur in one of the laws on which all those applications depend. Perhaps the most important service which science can render to practical affairs is to insist that laws can only be expected to hold between things which are themselves the expression of laws.

The last of the main sources of popular error is connected with a peculiar form of law which brevity has forbidden us to discuss hitherto. We have spoken of laws as asserting invariable associations. Now, a very slight acquaintance with science will suggest that this view is unduly narrow ; it may seem that some laws in almost all sciences (and almost all laws in some sciences) assert that one event is associated with another, not invariably, but usually or nearly always. Thus, if meteorology, the science of weather, has any laws at all, they would seem to be of this type ; nobody pretends that it is possible at present, or likely to be possible in the near future, to predict the weather exactly, especially a long time ahead ; the most we can hope for is to discover rules which will enable us generally to predict rightly. Another instance may be found in the study of heredity. It is an undoubted fact that, whether in plants, animals, or human beings, children of the same parents generally resemble each other and their parents more than they do others not closely related. But even the great progress in our knowledge of the laws of inheritance which has been made in recent years has not brought us near to a position in which it can be predicted (except in a few very simple cases) what exactly will be the property of each child of known parents. We know some rules, but they are not the exact and invariable rules which we have hitherto regarded as constituting laws.

The pure scientific view of such laws is very interesting. Put briefly it is that in such cases we have a mingling of two opposed agencies. There are laws concerned in such events, laws as strict and as invariable as those which are regarded as typical, but they are acting as it were on events governed not by law but by chance. The result of any law or set of laws depends (cf. p. 167) not only on the laws but on the events to which they are applied ; the irregularity that occurs in the study of the weather or of heredity is an irregularity in such events. Moreover, when science uses the conception of events governed by chance, it means something much more definite than is associated with that word by popular use. We say in ordinary discourse that an event is due to pure chance when we are completely ignorant whether it will or will not happen. But complete ignorance can never be a basis for knowledge, and the scientific conception of chance, which does lead to knowledge, implies only a certain limited degree of ignorance associated with a limited degree of knowledge. It is impossible here to discuss accurately what ignorance and what knowledge are implied, but it may be said roughly that the ignorance concerns each particular happening of the event, while the knowledge concerns a very long series of a great number of happenings. Thus, scientifically, it is an even chance whether a penny falls head or tail, because while we are perfectly ignorant whether on each particular toss it will fall head or tail, we are perfectly certain that in a sufficiently long series of tosses, heads and tails will be nearly equally distributed.

When, therefore, science finds, as in the study of weather or of heredity, phenomena which show some regularity, but not complete regularity, it tries to analyse them into perfectly regular laws acting on " chance " material. And the first step in the analysis is always to examine long series of the phenomena and to try to discover in these long series regularities which are not found in the individual members of them ; the regularity will usually consist in each of the alternative phenomena happening in a definite proportion of the trials. When such regularity has been found, a second step can sometimes be taken and an analysis made into strict laws applied to events which are governed by the particular form of regularity which science regards as *pure* chance. If that step can be taken, the scientific problem is solved, for *pure* chance, like strict law, is one of the ultimate conceptions of science. But there is often (as in meteorology) a long interval between the first step and the second, and in that interval all that is known is a regularity in the long series which is usually called a " statistical " law.

This procedure, like most scientific procedure, is borrowed and developed from common sense ; but— and this is the reason why it is mentioned here—it is here that modern common sense lags most behind scientific method. It is a familiar saying that statistics can prove anything ; and so they can in the hands of those not trained in scientific analysis. Statistical laws are one of the most abundant sources of popular fallacies which arise both from an ignorance of what such a law means and a still greater ignorance of how it is to be established. A statistical law does not state that something always happens, but that it happens more (or less) frequently than something else ; the quotation of instances of the thing happening are quite irrelevant to a proof of the law, unless there is at the same time a careful collection of instances in which it does not happen. Moreover, the clear distinction between a true law and a statistical law is not generally appreciated. A statistical law, which is really scientific, is made utterly fallacious in its application because it is interpreted as if it predicted the result of individual trials.

For reasons which have been given already, many of the laws that are most important in their practical application are statistical laws ; and anyone, with a little reflection, can suggest any number of examples of them. Those which are generally familiar are usually entirely fallacious (e.g. laws of weather and of heredity), and even those which are true are habitually misapplied. The general misunderstanding of statistical laws, renders them peculiarly liable, not only to their special fallacies, but to all those which we have discussed before. False theories and prejudices lead men to notice only those instances which are favourable to the law they want to establish ; they fail to see that, if it were a true law, a single contrary instance would be sufficient to disprove it, while, if it is a statistical law, favourable instances prove nothing, unless contrary instances are examined with equal care. They forget, too, that a statistical law can never be the whole truth ; it may, for the time being, represent all the truth we can attain ; but our efforts should never cease, until the full analysis has been effected, and the domain of strict law carefully separated from that of pure chance. The invention of that method of analysis, which leads to a possibility of prediction and control utterly impossible while knowledge is still in the statistical stage, is one of the things which makes science indispensable to the conduct of the affairs of practical life.

CONCLUSION

Our examination of the errors into which uninstructed reasoning is liable to mislead mankind, affected by so many prejudices and superstitions, shows immediately how and why science is indispensable, if any valuable lessons are to be drawn from the most ordinary experience. In the first place, science brings to the analysis of such experience the conception of definite, positive, and fixed law. For the vague conception of a law as a predominating, though variable, association, always liable to be distorted by circumstances, or even, like the laws of mankind, superseded by the vagaries of some higher authority, science substitutes, as its basic and most important conception, the association which is absolutely invariable, unchanging, universal. We may not always be successful in finding such laws, but our firm belief that they are to be found never wavers ; we never have the smallest reason to abandon our fundamental conviction that all events and changes, except in so far as they are the direct outcome of volition (and therefore immediately controllable), can be analysed into, and interpreted by, laws of the strictest form. It is only those who are guided by such a conviction that can hope to bring order and form into the infinite complexity of everyday experience.

However, such a conviction by itself would probably be of little avail. If we only knew from the outset that the analysis and explanation of experience could not be effected, except by disentangling from it the strict laws of science, every fresh problem would probably have to await solution until it came to the notice of some great genius ; for, as we noticed before, the discovery of a *wholly* new law is one of the greatest achievements of mankind. But we know much more ; the long series of laws which have been discovered, indicate where new laws are to be sought. We know that the terms involved in a new law must themselves be associated by a law. Moreover, the laws which define terms involved in other laws, though numerous, form a well-recognized class ; there are the laws defining various kinds of substances or various species of living beings, those defining forces, volumes, electric currents, the many forces of energy, and all the various measurable quantities of physics ; a complete list of them would fill a text-book, and yet their number is finite and comprehended by all serious students of the branches of science in which they are involved. Such students know that, when they try to analyse and explain new experience, it is between a definite class of terms that the necessary laws must be sought ; and that knowledge reduces the problem to one within the compass of a normal intellect, provided it is well trained. He who seeks to solve the problem without that knowledge and without the training on which it is based, cannot hope for even partial success, unless he can boast the powers of a Galileo or a Faraday.

And science provides yet another clue. It has established theories as well as laws. Its theories do not cover the full extent of its laws, and in some sciences little guidance except that of " empirical " laws is available. But where theories exist, they serve very closely to limit the laws by means which it is worth while to try to analyse experience ; no law contradictory of a firmly-rooted theory is worth examining till all other alternatives have been exhausted. Here uninstructed inquirers are at a still greater disadvantage, compared with those familiar with the results of science, for while many of the terms involved in scientific laws are vaguely familiar to every one, it is only those who have studied seriously, who have any knowledge of theories.

Here a word of warning should be given. " Theory " is always used in the book to mean the special class of propositions discussed in Chapter V. When in popular parlance " theory " is contrasted with " practice " it is often not this kind of theory that is meant at all. The plain man—I do not think this is an overstatement—calls a " theory " anything he does not understand, especially if the conclusions it is used to support are distasteful to him. Arguments about matters in which science is concerned, though they are denounced as wildly " theoretical," often depend on nothing but firmly-established law. It is only because he does not understand " theory " that the plain man is apt to compare it unfavourably with " practice," by which he means what he can understand. The idea that something can be " true in theory but false in practice " is due to mere ignorance ; if any portion of " practice," about which there is no doubt, is inconsistent with some " theory," then the " theory " (whether it is a law or what *we* call a theory) is false—and there is an end of it. But it may easily happen, and does often happen, that a " theory " is misinterpreted by those who fail to understand it ; it appears to predict something inconsistent with practice only because its real meaning is not grasped. It is certainly true that those who do not understand " theory " had better leave it alone ; reliance on misunderstood theory is certainly quite as dangerous as reliance on uninstructed " practice."

And here we come to the conclusion about the relation of science to everyday life, which it seems to me most important to enforce. Those unversed in the ways of science often regard it as a body of fixed knowledge contained in text-books and treatises, from which anyone who takes the trouble can extract all the information on any subject which science has to offer ; they think of it as something that can be learnt as the multiplication table can be learnt, and consider that anyone who has " done " science at school or college is complete master of its mysteries. Nothing could be further from the truth regarding science applied to practical problems. It is scarcely ever possible, even for the most learned student, to offer a complete and satisfying explanation of any difficulty, merely on the basis of established knowledge ; there is also some element in the problem which has not yet engaged scientific attention. Applied science, like pure science, is not a set of immutable principles and propositions ; it is rather an instrument of thought and a way of thinking. Every practical problem is really a problem in research, leading to the advancement of pure learning as well as to material efficiency ; indeed almost all the problems by the solution of which science has actually advanced have been suggested, more or less directly, by the familiar experiences of everyday life

This tremendous instrument of research can be mastered, this new way of thinking can be acquired, only by long training and laborious exercise. It is not, or it ought not to be, the academic student in the pure refined air of the laboratory who makes the knowledge, and the hard-handed and hard-headed worker who applies it to its needs. The man who can make new knowledge is the man, and the only man, to apply it.

Pure and applied science are the roots and the branches of the tree of experimental knowledge ; theory and practice are inseparably interwoven, and cannot be torn asunder without grave injury to both. The intellectual and the material health of society depend on the maintenance of their close connexion. A few years ago there was a tendency for true science to be confined to the laboratory, for its students to become thin-blooded, deprived of the invigorating air of industrial life, while industry wilted from neglect. To-day there are perhaps some signs of an extravagant reaction ; industrial science receives all the support and all the attention, while the universities, the nursing mothers of all science and all learning, are left to starve. The danger of rushing from one extreme to another will not be avoided until there is a general consciousness of what science means, both as a source of intellectual satisfaction and as a means of attaining material desires. We cannot all be— it is not desirable that we all should be—close students of science ; but we can all appreciate in some measure what are its aims, its methods, its uses. Science, like art, should not be something extraneous, added as a decoration to the other activities of existence ; it should be part of them, inspiring our most trivial actions as well as our noblest thoughts.

INDEX

A CATALOG OF SELECTED DOVER
BOOKS IN ALL FIELDS OF INTEREST

DRAWINGS OF REMBRANDT, edited by Seymour Slive. Updated Lippmann, Hofstede de Groot edition, with definitive scholarly apparatus. All portraits, biblical sketches, landscapes, nudes. Oriental figures, classical studies, together with selection of work by followers. 550 illustrations. Total of 630pp. 9⅛ × 12¼.
21485-0, 21486-9 Pa., Two-vol. set $25.00

GHOST AND HORROR STORIES OF AMBROSE BIERCE, Ambrose Bierce. 24 tales vividly imagined, strangely prophetic, and decades ahead of their time in technical skill: "The Damned Thing," "An Inhabitant of Carcosa," "The Eyes of the Panther," "Moxon's Master," and 20 more. 199pp. 5⅜ × 8½. 20767-6 Pa. $3.95

ETHICAL WRITINGS OF MAIMONIDES, Maimonides. Most significant ethical works of great medieval sage, newly translated for utmost precision, readability. Laws Concerning Character Traits, Eight Chapters, more. 192pp. 5⅜ × 8½.
24522-5 Pa. $4.50

THE EXPLORATION OF THE COLORADO RIVER AND ITS CANYONS, J. W. Powell. Full text of Powell's 1,000-mile expedition down the fabled Colorado in 1869. Superb account of terrain, geology, vegetation, Indians, famine, mutiny, treacherous rapids, mighty canyons, during exploration of last unknown part of continental U.S. 400pp. 5⅜ × 8½. 20094-9 Pa. $6.95

HISTORY OF PHILOSOPHY, Julián Marías. Clearest one-volume history on the market. Every major philosopher and dozens of others, to Existentialism and later. 505pp. 5⅜ × 8½. 21739-6 Pa. $8.50

ALL ABOUT LIGHTNING, Martin A. Uman. Highly readable non-technical survey of nature and causes of lightning, thunderstorms, ball lightning, St. Elmo's Fire, much more. Illustrated. 192pp. 5⅜ × 8½. 25237-X Pa. $5.95

SAILING ALONE AROUND THE WORLD, Captain Joshua Slocum. First man to sail around the world, alone, in small boat. One of great feats of seamanship told in delightful manner. 67 illustrations. 294pp. 5⅜ × 8½. 20326-3 Pa. $4.50

LETTERS AND NOTES ON THE MANNERS, CUSTOMS AND CONDITIONS OF THE NORTH AMERICAN INDIANS, George Catlin. Classic account of life among Plains Indians: ceremonies, hunt, warfare, etc. 312 plates. 572pp. of text. 6⅛ × 9¼. 22118-0, 22119-9 Pa. Two-vol. set $15.90

ALASKA: The Harriman Expedition, 1899, John Burroughs, John Muir, et al. Informative, engrossing accounts of two-month, 9,000-mile expedition. Native peoples, wildlife, forests, geography, salmon industry, glaciers, more. Profusely illustrated. 240 black-and-white line drawings. 124 black-and-white photographs. 3 maps. Index. 576pp. 5⅜ × 8½. 25109-8 Pa. $11.95

THE BOOK OF BEASTS: Being a Translation from a Latin Bestiary of the Twelfth Century, T. H. White. Wonderful catalog real and fanciful beasts: manticore, griffin, phoenix, amphivius, jaculus, many more. White's witty erudite commentary on scientific, historical aspects. Fascinating glimpse of medieval mind. Illustrated. 296pp. 5⅜ × 8¼. (Available in U.S. only) 24609-4 Pa. $5.95

FRANK LLOYD WRIGHT: ARCHITECTURE AND NATURE With 160 Illustrations, Donald Hoffmann. Profusely illustrated study of influence of nature—especially prairie—on Wright's designs for Fallingwater, Robie House, Guggenheim Museum, other masterpieces. 96pp. 9¼ × 10¾. 25098-9 Pa. $7.95

FRANK LLOYD WRIGHT'S FALLINGWATER, Donald Hoffmann. Wright's famous waterfall house: planning and construction of organic idea. History of site, owners, Wright's personal involvement. Photographs of various stages of building. Preface by Edgar Kaufmann, Jr. 100 illustrations. 112pp. 9¼ × 10.
23671-4 Pa. $7.95

YEARS WITH FRANK LLOYD WRIGHT: Apprentice to Genius, Edgar Tafel. Insightful memoir by a former apprentice presents a revealing portrait of Wright the man, the inspired teacher, the greatest American architect. 372 black-and-white illustrations. Preface. Index. vi + 228pp. 8¼ × 11. 24801-1 Pa. $9.95

THE STORY OF KING ARTHUR AND HIS KNIGHTS, Howard Pyle. Enchanting version of King Arthur fable has delighted generations with imaginative narratives of exciting adventures and unforgettable illustrations by the author. 41 illustrations. xviii + 313pp. 6⅛ × 9¼. 21445-1 Pa. $5.95

THE GODS OF THE EGYPTIANS, E. A. Wallis Budge. Thorough coverage of numerous gods of ancient Egypt by foremost Egyptologist. Information on evolution of cults, rites and gods; the cult of Osiris; the Book of the Dead and its rites; the sacred animals and birds; Heaven and Hell; and more. 956pp. 6⅛ × 9¼.
22055-9, 22056-7 Pa., Two-vol. set $20.00

A THEOLOGICO-POLITICAL TREATISE, Benedict Spinoza. Also contains unfinished *Political Treatise*. Great classic on religious liberty, theory of government on common consent. R. Elwes translation. Total of 421pp. 5⅜ × 8½.
20249-6 Pa. $6.95

INCIDENTS OF TRAVEL IN CENTRAL AMERICA, CHIAPAS, AND YUCATAN, John L. Stephens. Almost single-handed discovery of Maya culture; exploration of ruined cities, monuments, temples; customs of Indians. 115 drawings. 892pp. 5⅜ × 8½. 22404-X, 22405-8 Pa., Two-vol. set $15.90

LOS CAPRICHOS, Francisco Goya. 80 plates of wild, grotesque monsters and caricatures. Prado manuscript included. 183pp. 6⅜ × 9⅜. 22384-1 Pa. $4.95

AUTOBIOGRAPHY: The Story of My Experiments with Truth, Mohandas K. Gandhi. Not hagiography, but Gandhi in his own words. Boyhood, legal studies, purification, the growth of the Satyagraha (nonviolent protest) movement. Critical, inspiring work of the man who freed India. 480pp. 5⅜ × 8½. (Available in U.S. only)
24593-4 Pa. $6.95

ILLUSTRATED DICTIONARY OF HISTORIC ARCHITECTURE, edited by Cyril M. Harris. Extraordinary compendium of clear, concise definitions for over 5,000 important architectural terms complemented by over 2,000 line drawings. Covers full spectrum of architecture from ancient ruins to 20th-century Modernism. Preface. 592pp. 7½ × 9⅜. 24444-X Pa. $14.95

THE NIGHT BEFORE CHRISTMAS, Clement Moore. Full text, and woodcuts from original 1848 book. Also critical, historical material. 19 illustrations. 40pp. 4⅝ × 6. 22797-9 Pa. $2.25

THE LESSON OF JAPANESE ARCHITECTURE: 165 Photographs, Jiro Harada. Memorable gallery of 165 photographs taken in the 1930's of exquisite Japanese homes of the well-to-do and historic buildings. 13 line diagrams. 192pp. 8⅜ × 11¼. 24778-3 Pa. $8.95

THE AUTOBIOGRAPHY OF CHARLES DARWIN AND SELECTED LETTERS, edited by Francis Darwin. The fascinating life of eccentric genius composed of an intimate memoir by Darwin (intended for his children); commentary by his son, Francis; hundreds of fragments from notebooks, journals, papers; and letters to and from Lyell, Hooker, Huxley, Wallace and Henslow. xi + 365pp. 5⅜ × 8. 20479-0 Pa. $5.95

WONDERS OF THE SKY: Observing Rainbows, Comets, Eclipses, the Stars and Other Phenomena, Fred Schaaf. Charming, easy-to-read poetic guide to all manner of celestial events visible to the naked eye. Mock suns, glories, Belt of Venus, more. Illustrated. 299pp. 5¼ × 8¼. 24402-4 Pa. $7.95

BURNHAM'S CELESTIAL HANDBOOK, Robert Burnham, Jr. Thorough guide to the stars beyond our solar system. Exhaustive treatment. Alphabetical by constellation: Andromeda to Cetus in Vol. 1; Chamaeleon to Orion in Vol. 2; and Pavo to Vulpecula in Vol. 3. Hundreds of illustrations. Index in Vol. 3. 2,000pp. 6⅛ × 9¼. 23567-X, 23568-8, 23673-0 Pa., Three-vol. set $36.85

STAR NAMES: Their Lore and Meaning, Richard Hinckley Allen. Fascinating history of names various cultures have given to constellations and literary and folkloristic uses that have been made of stars. Indexes to subjects. Arabic and Greek names. Biblical references. Bibliography. 563pp. 5⅜ × 8½. 21079-0 Pa. $7.95

THIRTY YEARS THAT SHOOK PHYSICS: The Story of Quantum Theory, George Gamow. Lucid, accessible introduction to influential theory of energy and matter. Careful explanations of Dirac's anti-particles, Bohr's model of the atom, much more. 12 plates. Numerous drawings. 240pp. 5⅜ × 8½. 24895-X Pa. $4.95

CHINESE DOMESTIC FURNITURE IN PHOTOGRAPHS AND MEASURED DRAWINGS, Gustav Ecke. A rare volume, now affordably priced for antique collectors, furniture buffs and art historians. Detailed review of styles ranging from early Shang to late Ming. Unabridged republication. 161 black-and-white drawings, photos. Total of 224pp. 8⅜ × 11¼. (Available in U.S. only) 25171-3 Pa. $12.95

VINCENT VAN GOGH: A Biography, Julius Meier-Graefe. Dynamic, penetrating study of artist's life, relationship with brother, Theo, painting techniques, travels, more. Readable, engrossing. 160pp. 5⅜ × 8½. (Available in U.S. only) 25253-1 Pa. $3.95

HOW TO WRITE, Gertrude Stein. Gertrude Stein claimed anyone could understand her unconventional writing—here are clues to help. Fascinating improvisations, language experiments, explanations illuminate Stein's craft and the art of writing. Total of 414pp. 4⅝ × 6⅜. 23144-5 Pa. $5.95

ADVENTURES AT SEA IN THE GREAT AGE OF SAIL: Five Firsthand Narratives, edited by Elliot Snow. Rare true accounts of exploration, whaling, shipwreck, fierce natives, trade, shipboard life, more. 33 illustrations. Introduction. 353pp. 5⅜ × 8½. 25177-2 Pa. $7.95

THE HERBAL OR GENERAL HISTORY OF PLANTS, John Gerard. Classic descriptions of about 2,850 plants—with over 2,700 illustrations—includes Latin and English names, physical descriptions, varieties, time and place of growth, more. 2,706 illustrations. xlv + 1,678pp. 8½ × 12¼. 23147-X Cloth. $75.00

DOROTHY AND THE WIZARD IN OZ, L. Frank Baum. Dorothy and the Wizard visit the center of the Earth, where people are vegetables, glass houses grow and Oz characters reappear. Classic sequel to Wizard of Oz. 256pp. 5⅜ × 8.
24714-7 Pa. $4.95

SONGS OF EXPERIENCE: Facsimile Reproduction with 26 Plates in Full Color, William Blake. This facsimile of Blake's original "Illuminated Book" reproduces 26 full-color plates from a rare 1826 edition. Includes "The Tyger," "London," "Holy Thursday," and other immortal poems. 26 color plates. Printed text of poems. 48pp. 5¼ × 7. 24636-1 Pa. $3.50

SONGS OF INNOCENCE, William Blake. The first and most popular of Blake's famous "Illuminated Books," in a facsimile edition reproducing all 31 brightly colored plates. Additional printed text of each poem. 64pp. 5¼ × 7.
22764-2 Pa. $3.50

PRECIOUS STONES, Max Bauer. Classic, thorough study of diamonds, rubies, emeralds, garnets, etc.: physical character, occurrence, properties, use, similar topics. 20 plates, 8 in color. 94 figures. 659pp. 6⅛ × 9¼.
21910-0, 21911-9 Pa., Two-vol. set $14.90

ENCYCLOPEDIA OF VICTORIAN NEEDLEWORK, S. F. A. Caulfeild and Blanche Saward. Full, precise descriptions of stitches, techniques for dozens of needlecrafts—most exhaustive reference of its kind. Over 800 figures. Total of 679pp. 8⅛ × 11. Two volumes. Vol. 1 22800-2 Pa. $10.95
Vol. 2 22801-0 Pa. $10.95

THE MARVELOUS LAND OF OZ, L. Frank Baum. Second Oz book, the Scarecrow and Tin Woodman are back with hero named Tip, Oz magic. 136 illustrations. 287pp. 5⅜ × 8½. 20692-0 Pa. $5.95

WILD FOWL DECOYS, Joel Barber. Basic book on the subject, by foremost authority and collector. Reveals history of decoy making and rigging, place in American culture, different kinds of decoys, how to make them, and how to use them. 140 plates. 156pp. 7⅞ × 10¾. 20011-6 Pa. $7.95

HISTORY OF LACE, Mrs. Bury Palliser. Definitive, profusely illustrated chronicle of lace from earliest times to late 19th century. Laces of Italy, Greece, England, France, Belgium, etc. Landmark of needlework scholarship. 266 illustrations. 672pp. 6⅛ × 9¼. 24742-2 Pa. $14.95

ILLUSTRATED GUIDE TO SHAKER FURNITURE, Robert Meader. All furniture and appurtenances, with much on unknown local styles. 235 photos. 146pp. 9 × 12. 22819-3 Pa. $7.95

WHALE SHIPS AND WHALING: A Pictorial Survey, George Francis Dow. Over 200 vintage engravings, drawings, photographs of barks, brigs, cutters, other vessels. Also harpoons, lances, whaling guns, many other artifacts. Comprehensive text by foremost authority. 207 black-and-white illustrations. 288pp. 6 × 9. 24808-9 Pa. $8.95

THE BERTRAMS, Anthony Trollope. Powerful portrayal of blind self-will and thwarted ambition includes one of Trollope's most heartrending love stories. 497pp. 5⅜ × 8½. 25119-5 Pa. $8.95

ADVENTURES WITH A HAND LENS, Richard Headstrom. Clearly written guide to observing and studying flowers and grasses, fish scales, moth and insect wings, egg cases, buds, feathers, seeds, leaf scars, moss, molds, ferns, common crystals, etc.—all with an ordinary, inexpensive magnifying glass. 209 exact line drawings aid in your discoveries. 220pp. 5⅜ × 8½. 23330-8 Pa. $3.95

RODIN ON ART AND ARTISTS, Auguste Rodin. Great sculptor's candid, wide-ranging comments on meaning of art; great artists; relation of sculpture to poetry, painting, music; philosophy of life, more. 76 superb black-and-white illustrations of Rodin's sculpture, drawings and prints. 119pp. 8⅜ × 11¼. 24487-3 Pa. $6.95

FIFTY CLASSIC FRENCH FILMS, 1912–1982: A Pictorial Record, Anthony Slide. Memorable stills from Grand Illusion, Beauty and the Beast, Hiroshima, Mon Amour, many more. Credits, plot synopses, reviews, etc. 160pp. 8¼ × 11. 25256-6 Pa. $11.95

THE PRINCIPLES OF PSYCHOLOGY, William James. Famous long course complete, unabridged. Stream of thought, time perception, memory, experimental methods; great work decades ahead of its time. 94 figures. 1,391pp. 5⅜ × 8½. 20381-6, 20382-4 Pa., Two-vol. set $19.90

BODIES IN A BOOKSHOP, R. T. Campbell. Challenging mystery of blackmail and murder with ingenious plot and superbly drawn characters. In the best tradition of British suspense fiction. 192pp. 5⅜ × 8½. 24720-1 Pa. $3.95

CALLAS: PORTRAIT OF A PRIMA DONNA, George Jellinek. Renowned commentator on the musical scene chronicles incredible career and life of the most controversial, fascinating, influential operatic personality of our time. 64 black-and-white photographs. 416pp. 5⅜ × 8¼. 25047-4 Pa. $7.95

GEOMETRY, RELATIVITY AND THE FOURTH DIMENSION, Rudolph Rucker. Exposition of fourth dimension, concepts of relativity as Flatland characters continue adventures. Popular, easily followed yet accurate, profound. 141 illustrations. 133pp. 5⅜ × 8½. 23400-2 Pa. $3.50

HOUSEHOLD STORIES BY THE BROTHERS GRIMM, with pictures by Walter Crane. 53 classic stories—Rumpelstiltskin, Rapunzel, Hansel and Gretel, the Fisherman and his Wife, Snow White, Tom Thumb, Sleeping Beauty, Cinderella, and so much more—lavishly illustrated with original 19th century drawings. 114 illustrations. x + 269pp. 5⅜ × 8½. 21080-4 Pa. $4.50

SUNDIALS, Albert Waugh. Far and away the best, most thorough coverage of ideas, mathematics concerned, types, construction, adjusting anywhere. Over 100 illustrations. 230pp. 5⅜ × 8½. 22947-5 Pa. $4.00

PICTURE HISTORY OF THE NORMANDIE: With 190 Illustrations, Frank O. Braynard. Full story of legendary French ocean liner: Art Deco interiors, design innovations, furnishings, celebrities, maiden voyage, tragic fire, much more. Extensive text. 144pp. 8⅞ × 11¼. 25257-4 Pa. $9.95

THE FIRST AMERICAN COOKBOOK: A Facsimile of "American Cookery," 1796, Amelia Simmons. Facsimile of the first American-written cookbook published in the United States contains authentic recipes for colonial favorites—pumpkin pudding, winter squash pudding, spruce beer, Indian slapjacks, and more. Introductory Essay and Glossary of colonial cooking terms. 80pp. 5⅜ × 8½. 24710-4 Pa. $3.50

101 PUZZLES IN THOUGHT AND LOGIC, C. R. Wylie, Jr. Solve murders and robberies, find out which fishermen are liars, how a blind man could possibly identify a color—purely by your own reasoning! 107pp. 5⅜ × 8½. 20367-0 Pa. $2.00

THE BOOK OF WORLD-FAMOUS MUSIC—CLASSICAL, POPULAR AND FOLK, James J. Fuld. Revised and enlarged republication of landmark work in musico-bibliography. Full information about nearly 1,000 songs and compositions including first lines of music and lyrics. New supplement. Index. 800pp. 5⅜ × 8¼. 24857-7 Pa. $14.95

ANTHROPOLOGY AND MODERN LIFE, Franz Boas. Great anthropologist's classic treatise on race and culture. Introduction by Ruth Bunzel. Only inexpensive paperback edition. 255pp. 5⅜ × 8½. 25245-0 Pa. $5.95

THE TALE OF PETER RABBIT, Beatrix Potter. The inimitable Peter's terrifying adventure in Mr. McGregor's garden, with all 27 wonderful, full-color Potter illustrations. 55pp. 4¼ × 5½. (Available in U.S. only) 22827-4 Pa. $1.75

THREE PROPHETIC SCIENCE FICTION NOVELS, H. G. Wells. *When the Sleeper Wakes, A Story of the Days to Come* and *The Time Machine* (full version). 335pp. 5⅜ × 8½. (Available in U.S. only) 20605-X Pa. $5.95

APICIUS COOKERY AND DINING IN IMPERIAL ROME, edited and translated by Joseph Dommers Vehling. Oldest known cookbook in existence offers readers a clear picture of what foods Romans ate, how they prepared them, etc. 49 illustrations. 301pp. 6⅛ × 9¼. 23563-7 Pa. $6.00

SHAKESPEARE LEXICON AND QUOTATION DICTIONARY, Alexander Schmidt. Full definitions, locations, shades of meaning of every word in plays and poems. More than 50,000 exact quotations. 1,485pp. 6½ × 9¼. 22726-X, 22727-8 Pa., Two-vol. set $27.90

THE WORLD'S GREAT SPEECHES, edited by Lewis Copeland and Lawrence W. Lamm. Vast collection of 278 speeches from Greeks to 1970. Powerful and effective models; unique look at history. 842pp. 5⅜ × 8½. 20468-5 Pa. $10.95

THE BLUE FAIRY BOOK, Andrew Lang. The first, most famous collection, with many familiar tales: Little Red Riding Hood, Aladdin and the Wonderful Lamp, Puss in Boots, Sleeping Beauty, Hansel and Gretel, Rumpelstiltskin; 37 in all. 138 illustrations. 390pp. 5⅜ × 8½. 21437-0 Pa. $5.95

THE STORY OF THE CHAMPIONS OF THE ROUND TABLE, Howard Pyle. Sir Launcelot, Sir Tristram and Sir Percival in spirited adventures of love and triumph retold in Pyle's inimitable style. 50 drawings, 31 full-page. xviii + 329pp. 6½ × 9¼. 21883-X Pa. $6.95

AUDUBON AND HIS JOURNALS, Maria Audubon. Unmatched two-volume portrait of the great artist, naturalist and author contains his journals, an excellent biography by his granddaughter, expert annotations by the noted ornithologist, Dr. Elliott Coues, and 37 superb illustrations. Total of 1,200pp. 5⅜ × 8.
Vol. I 25143-8 Pa. $8.95
Vol. II 25144-6 Pa. $8.95

GREAT DINOSAUR HUNTERS AND THEIR DISCOVERIES, Edwin H. Colbert. Fascinating, lavishly illustrated chronicle of dinosaur research, 1820's to 1960. Achievements of Cope, Marsh, Brown, Buckland, Mantell, Huxley, many others. 384pp. 5¼ × 8¼. 24701-5 Pa. $6.95

THE TASTEMAKERS, Russell Lynes. Informal, illustrated social history of American taste 1850's–1950's. First popularized categories Highbrow, Lowbrow, Middlebrow. 129 illustrations. New (1979) afterword. 384pp. 6 × 9.
23993-4 Pa. $6.95

DOUBLE CROSS PURPOSES, Ronald A. Knox. A treasure hunt in the Scottish Highlands, an old map, unidentified corpse, surprise discoveries keep reader guessing in this cleverly intricate tale of financial skullduggery. 2 black-and-white maps. 320pp. 5⅜ × 8½. (Available in U.S. only) 25032-6 Pa. $5.95

AUTHENTIC VICTORIAN DECORATION AND ORNAMENTATION IN FULL COLOR: 46 Plates from "Studies in Design," Christopher Dresser. Superb full-color lithographs reproduced from rare original portfolio of a major Victorian designer. 48pp. 9¼ × 12¼. 25083-0 Pa. $7.95

PRIMITIVE ART, Franz Boas. Remains the best text ever prepared on subject, thoroughly discussing Indian, African, Asian, Australian, and, especially, Northern American primitive art. Over 950 illustrations show ceramics, masks, totem poles, weapons, textiles, paintings, much more. 376pp. 5⅜ × 8. 20025-6 Pa. $6.95

SIDELIGHTS ON RELATIVITY, Albert Einstein. Unabridged republication of two lectures delivered by the great physicist in 1920–21. *Ether and Relativity* and *Geometry and Experience*. Elegant ideas in non-mathematical form, accessible to intelligent layman. vi + 56pp. 5⅜ × 8½. 24511-X Pa. $2.95

THE WIT AND HUMOR OF OSCAR WILDE, edited by Alvin Redman. More than 1,000 ripostes, paradoxes, wisecracks: Work is the curse of the drinking classes, I can resist everything except temptation, etc. 258pp. 5⅜ × 8½. 20602-5 Pa. $3.95

ADVENTURES WITH A MICROSCOPE, Richard Headstrom. 59 adventures with clothing fibers, protozoa, ferns and lichens, roots and leaves, much more. 142 illustrations. 232pp. 5⅜ × 8½. 23471-1 Pa. $3.95

PLANTS OF THE BIBLE, Harold N. Moldenke and Alma L. Moldenke. Standard reference to all 230 plants mentioned in Scriptures. Latin name, biblical reference, uses, modern identity, much more. Unsurpassed encyclopedic resource for scholars, botanists, nature lovers, students of Bible. Bibliography. Indexes. 123 black-and-white illustrations. 384pp. 6 × 9. 25069-5 Pa. $8.95

FAMOUS AMERICAN WOMEN: A Biographical Dictionary from Colonial Times to the Present, Robert McHenry, ed. From Pocahontas to Rosa Parks, 1,035 distinguished American women documented in separate biographical entries. Accurate, up-to-date data, numerous categories, spans 400 years. Indices. 493pp. 6½ × 9¼. 24523-3 Pa. $9.95

THE FABULOUS INTERIORS OF THE GREAT OCEAN LINERS IN HISTORIC PHOTOGRAPHS, William H. Miller, Jr. Some 200 superb photographs capture exquisite interiors of world's great "floating palaces"—1890's to 1980's: *Titanic, Ile de France, Queen Elizabeth, United States, Europa,* more. Approx. 200 black-and-white photographs. Captions. Text. Introduction. 160pp. 8⅞ × 11¾. 24756-2 Pa. $9.95

THE GREAT LUXURY LINERS, 1927-1954: A Photographic Record, William H. Miller, Jr. Nostalgic tribute to heyday of ocean liners. 186 photos of Ile de France, Normandie, Leviathan, Queen Elizabeth, United States, many others. Interior and exterior views. Introduction. Captions. 160pp. 9 × 12. 24056-8 Pa. $9.95

A NATURAL HISTORY OF THE DUCKS, John Charles Phillips. Great landmark of ornithology offers complete detailed coverage of nearly 200 species and subspecies of ducks: gadwall, sheldrake, merganser, pintail, many more. 74 full-color plates, 102 black-and-white. Bibliography. Total of 1,920pp. 8⅝ × 11¼. 25141-1, 25142-X Cloth. Two-vol. set $100.00

THE SEAWEED HANDBOOK: An Illustrated Guide to Seaweeds from North Carolina to Canada, Thomas F. Lee. Concise reference covers 78 species. Scientific and common names, habitat, distribution, more. Finding keys for easy identification. 224pp. 5⅜ × 8½. 25215-9 Pa. $5.95

THE TEN BOOKS OF ARCHITECTURE: The 1755 Leoni Edition, Leon Battista Alberti. Rare classic helped introduce the glories of ancient architecture to the Renaissance. 68 black-and-white plates. 336pp. 8⅜ × 11¼. 25239-6 Pa. $14.95

MISS MACKENZIE, Anthony Trollope. Minor masterpieces by Victorian master unmasks many truths about life in 19th-century England. First inexpensive edition in years. 392pp. 5⅜ × 8½. 25201-9 Pa. $7.95

THE RIME OF THE ANCIENT MARINER, Gustave Doré, Samuel Taylor Coleridge. Dramatic engravings considered by many to be his greatest work. The terrifying space of the open sea, the storms and whirlpools of an unknown ocean, the ice of Antarctica, more—all rendered in a powerful, chilling manner. Full text. 38 plates. 77pp. 9¼ × 12. 22305-1 Pa. $4.95

THE EXPEDITIONS OF ZEBULON MONTGOMERY PIKE, Zebulon Montgomery Pike. Fascinating first-hand accounts (1805-6) of exploration of Mississippi River, Indian wars, capture by Spanish dragoons, much more. 1,088pp. 5⅜ × 8½. 25254-X, 25255-8 Pa. Two-vol. set $23.90

CATALOG OF DOVER BOOKS

A CONCISE HISTORY OF PHOTOGRAPHY: Third Revised Edition, Helmut Gernsheim. Best one-volume history—camera obscura, photochemistry, daguerreotypes, evolution of cameras, film, more. Also artistic aspects—landscape, portraits, fine art, etc. 281 black-and-white photographs. 26 in color. 176pp. 8⅜ × 11¼. 25128-4 Pa. $12.95

THE DORÉ BIBLE ILLUSTRATIONS, Gustave Doré. 241 detailed plates from the Bible: the Creation scenes, Adam and Eve, Flood, Babylon, battle sequences, life of Jesus, etc. Each plate is accompanied by the verses from the King James version of the Bible. 241pp. 9 × 12. 23004-X Pa. $8.95

HUGGER-MUGGER IN THE LOUVRE, Elliot Paul. Second Homer Evans mystery-comedy. Theft at the Louvre involves sleuth in hilarious, madcap caper. "A knockout."—Books. 336pp. 5⅜ × 8½. 25185-3 Pa. $5.95

FLATLAND, E. A. Abbott. Intriguing and enormously popular science-fiction classic explores the complexities of trying to survive as a two-dimensional being in a three-dimensional world. Amusingly illustrated by the author. 16 illustrations. 103pp. 5⅜ × 8½. 20001-9 Pa. $2.00

THE HISTORY OF THE LEWIS AND CLARK EXPEDITION, Meriwether Lewis and William Clark, edited by Elliott Coues. Classic edition of Lewis and Clark's day-by-day journals that later became the basis for U.S. claims to Oregon and the West. Accurate and invaluable geographical, botanical, biological, meteorological and anthropological material. Total of 1,508pp. 5⅜ × 8½. 21268-8, 21269-6, 21270-X Pa. Three-vol. set $25.50

LANGUAGE, TRUTH AND LOGIC, Alfred J. Ayer. Famous, clear introduction to Vienna, Cambridge schools of Logical Positivism. Role of philosophy, elimination of metaphysics, nature of analysis, etc. 160pp. 5⅜ × 8½. (Available in U.S. and Canada only) 20010-8 Pa. $2.95

MATHEMATICS FOR THE NONMATHEMATICIAN, Morris Kline. Detailed, college-level treatment of mathematics in cultural and historical context, with numerous exercises. For liberal arts students. Preface. Recommended Reading Lists. Tables. Index. Numerous black-and-white figures. xvi + 641pp. 5⅜ × 8½. 24823-2 Pa. $11.95

28 SCIENCE FICTION STORIES, H. G. Wells. Novels, *Star Begotten* and *Men Like Gods*, plus 26 short stories: "Empire of the Ants," "A Story of the Stone Age," "The Stolen Bacillus," "In the Abyss," etc. 915pp. 5⅜ × 8½. (Available in U.S. only) 20265-8 Cloth. $10.95

HANDBOOK OF PICTORIAL SYMBOLS, Rudolph Modley. 3,250 signs and symbols, many systems in full; official or heavy commercial use. Arranged by subject. Most in Pictorial Archive series. 143pp. 8⅛ × 11. 23357-X Pa. $5.95

INCIDENTS OF TRAVEL IN YUCATAN, John L. Stephens. Classic (1843) exploration of jungles of Yucatan, looking for evidences of Maya civilization. Travel adventures, Mexican and Indian culture, etc. Total of 669pp. 5⅜ × 8½. 20926-1, 20927-X Pa., Two-vol. set $9.90

DEGAS: An Intimate Portrait, Ambroise Vollard. Charming, anecdotal memoir by famous art dealer of one of the greatest 19th-century French painters. 14 black-and-white illustrations. Introduction by Harold L. Van Doren. 96pp. 5⅜ × 8½.
25131-4 Pa. $3.95

PERSONAL NARRATIVE OF A PILGRIMAGE TO ALMANDINAH AND MECCAH, Richard Burton. Great travel classic by remarkably colorful personality. Burton, disguised as a Moroccan, visited sacred shrines of Islam, narrowly escaping death. 47 illustrations. 959pp. 5⅜ × 8½.   21217-3, 21218-1 Pa., Two-vol. set $17.90

PHRASE AND WORD ORIGINS, A. H. Holt. Entertaining, reliable, modern study of more than 1,200 colorful words, phrases, origins and histories. Much unexpected information. 254pp. 5⅜ × 8½.   20758-7 Pa. $4.95

THE RED THUMB MARK, R. Austin Freeman. In this first Dr. Thorndyke case, the great scientific detective draws fascinating conclusions from the nature of a single fingerprint. Exciting story, authentic science. 320pp. 5⅜ × 8½. (Available in U.S. only)   25210-8 Pa. $5.95

AN EGYPTIAN HIEROGLYPHIC DICTIONARY, E. A. Wallis Budge. Monumental work containing about 25,000 words or terms that occur in texts ranging from 3000 B.C. to 600 A.D. Each entry consists of a transliteration of the word, the word in hieroglyphs, and the meaning in English. 1,314pp. 6⅜ × 10.
23615-3, 23616-1 Pa., Two-vol. set $27.90

THE COMPLEAT STRATEGYST: Being a Primer on the Theory of Games of Strategy, J. D. Williams. Highly entertaining classic describes, with many illustrated examples, how to select best strategies in conflict situations. Prefaces. Appendices. xvi + 268pp. 5⅜ × 8½.   25101-2 Pa. $5.95

THE ROAD TO OZ, L. Frank Baum. Dorothy meets the Shaggy Man, little Button-Bright and the Rainbow's beautiful daughter in this delightful trip to the magical Land of Oz. 272pp. 5⅜ × 8.   25208-6 Pa. $4.95

POINT AND LINE TO PLANE, Wassily Kandinsky. Seminal exposition of role of point, line, other elements in non-objective painting. Essential to understanding 20th-century art. 127 illustrations. 192pp. 6½ × 9¼.   23808-3 Pa. $4.50

LADY ANNA, Anthony Trollope. Moving chronicle of Countess Lovel's bitter struggle to win for herself and daughter Anna their rightful rank and fortune—perhaps at cost of sanity itself. 384pp. 5⅜ × 8½.   24669-8 Pa. $6.95

EGYPTIAN MAGIC, E. A. Wallis Budge. Sums up all that is known about magic in Ancient Egypt: the role of magic in controlling the gods, powerful amulets that warded off evil spirits, scarabs of immortality, use of wax images, formulas and spells, the secret name, much more. 253pp. 5⅜ × 8½.   22681-6 Pa. $4.00

THE DANCE OF SIVA, Ananda Coomaraswamy. Preeminent authority unfolds the vast metaphysic of India: the revelation of her art, conception of the universe, social organization, etc. 27 reproductions of art masterpieces. 192pp. 5⅜ × 8½.
24817-8 Pa. $5.95

CHRISTMAS CUSTOMS AND TRADITIONS, Clement A. Miles. Origin, evolution, significance of religious, secular practices. Caroling, gifts, yule logs, much more. Full, scholarly yet fascinating; non-sectarian. 400pp. 5⅜ × 8½.
23354-5 Pa. $6.50

THE HUMAN FIGURE IN MOTION, Eadweard Muybridge. More than 4,500 stopped-action photos, in action series, showing undraped men, women, children jumping, lying down, throwing, sitting, wrestling, carrying, etc. 390pp. 7⅞ × 10⅝.
20204-6 Cloth. $19.95

THE MAN WHO WAS THURSDAY, Gilbert Keith Chesterton. Witty, fast-paced novel about a club of anarchists in turn-of-the-century London. Brilliant social, religious, philosophical speculations. 128pp. 5⅜ × 8½. 25121-7 Pa. $3.95

A CEZANNE SKETCHBOOK: Figures, Portraits, Landscapes and Still Lifes, Paul Cezanne. Great artist experiments with tonal effects, light, mass, other qualities in over 100 drawings. A revealing view of developing master painter, precursor of Cubism. 102 black-and-white illustrations. 144pp. 8¾ × 6⅜. 24790-2 Pa. $5.95

AN ENCYCLOPEDIA OF BATTLES: Accounts of Over 1,560 Battles from 1479 B.C. to the Present, David Eggenberger. Presents essential details of every major battle in recorded history, from the first battle of Megiddo in 1479 B.C. to Grenada in 1984. List of Battle Maps. New Appendix covering the years 1967–1984. Index. 99 illustrations. 544pp. 6½ × 9¼. 24913-1 Pa. $14.95

AN ETYMOLOGICAL DICTIONARY OF MODERN ENGLISH, Ernest Weekley. Richest, fullest work, by foremost British lexicographer. Detailed word histories. Inexhaustible. Total of 856pp. 6½ × 9¼.
21873-2, 21874-0 Pa., Two-vol. set $17.00

WEBSTER'S AMERICAN MILITARY BIOGRAPHIES, edited by Robert McHenry. Over 1,000 figures who shaped 3 centuries of American military history. Detailed biographies of Nathan Hale, Douglas MacArthur, Mary Hallaren, others. Chronologies of engagements, more. Introduction. Addenda. 1,033 entries in alphabetical order. xi + 548pp. 6½ × 9¼. (Available in U.S. only)
24758-9 Pa. $11.95

LIFE IN ANCIENT EGYPT, Adolf Erman. Detailed older account, with much not in more recent books: domestic life, religion, magic, medicine, commerce, and whatever else needed for complete picture. Many illustrations. 597pp. 5⅜ × 8½.
22632-8 Pa. $8.50

HISTORIC COSTUME IN PICTURES, Braun & Schneider. Over 1,450 costumed figures shown, covering a wide variety of peoples: kings, emperors, nobles, priests, servants, soldiers, scholars, townsfolk, peasants, merchants, courtiers, cavaliers, and more. 256pp. 8⅜ × 11¼. 23150-X Pa. $7.95

THE NOTEBOOKS OF LEONARDO DA VINCI, edited by J. P. Richter. Extracts from manuscripts reveal great genius; on painting, sculpture, anatomy, sciences, geography, etc. Both Italian and English. 186 ms. pages reproduced, plus 500 additional drawings, including studies for *Last Supper, Sforza* monument, etc. 860pp. 7⅞ × 10¾. (Available in U.S. only) 22572-0, 22573-9 Pa., Two-vol. set $25.90

THE ART NOUVEAU STYLE BOOK OF ALPHONSE MUCHA: All 72 Plates from "Documents Decoratifs" in Original Color, Alphonse Mucha. Rare copyright-free design portfolio by high priest of Art Nouveau. Jewelry, wallpaper, stained glass, furniture, figure studies, plant and animal motifs, etc. Only complete one-volume edition. 80pp. 9⅜ × 12¼. 24044-4 Pa. $8.95

ANIMALS: 1,419 COPYRIGHT-FREE ILLUSTRATIONS OF MAMMALS, BIRDS, FISH, INSECTS, ETC., edited by Jim Harter. Clear wood engravings present, in extremely lifelike poses, over 1,000 species of animals. One of the most extensive pictorial sourcebooks of its kind. Captions. Index. 284pp. 9 × 12. 23766-4 Pa. $9.95

OBELISTS FLY HIGH, C. Daly King. Masterpiece of American detective fiction, long out of print, involves murder on a 1935 transcontinental flight—"a very thrilling story"—NY Times. Unabridged and unaltered republication of the edition published by William Collins Sons & Co. Ltd., London, 1935. 288pp. 5⅜ × 8½. (Available in U.S. only) 25036-9 Pa. $4.95

VICTORIAN AND EDWARDIAN FASHION: A Photographic Survey, Alison Gernsheim. First fashion history completely illustrated by contemporary photographs. Full text plus 235 photos, 1840–1914, in which many celebrities appear. 240pp. 6½ × 9¼. 24205-6 Pa. $6.00

THE ART OF THE FRENCH ILLUSTRATED BOOK, 1700–1914, Gordon N. Ray. Over 630 superb book illustrations by Fragonard, Delacroix, Daumier, Doré, Grandville, Manet, Mucha, Steinlen, Toulouse-Lautrec and many others. Preface. Introduction. 633 halftones. Indices of artists, authors & titles, binders and provenances. Appendices. Bibliography. 608pp. 8⅜ × 11¼. 25086-5 Pa. $24.95

THE WONDERFUL WIZARD OF OZ, L. Frank Baum. Facsimile in full color of America's finest children's classic. 143 illustrations by W. W. Denslow. 267pp. 5⅜ × 8½. 20691-2 Pa. $5.95

FRONTIERS OF MODERN PHYSICS: New Perspectives on Cosmology, Relativity, Black Holes and Extraterrestrial Intelligence, Tony Rothman, et al. For the intelligent layman. Subjects include: cosmological models of the universe; black holes; the neutrino; the search for extraterrestrial intelligence. Introduction. 46 black-and-white illustrations. 192pp. 5⅜ × 8½. 24587-X Pa. $6.95

THE FRIENDLY STARS, Martha Evans Martin & Donald Howard Menzel. Classic text marshalls the stars together in an engaging, non-technical survey, presenting them as sources of beauty in night sky. 23 illustrations. Foreword. 2 star charts. Index. 147pp. 5⅜ × 8½. 21099-5 Pa. $3.50

FADS AND FALLACIES IN THE NAME OF SCIENCE, Martin Gardner. Fair, witty appraisal of cranks, quacks, and quackeries of science and pseudoscience: hollow earth, Velikovsky, orgone energy, Dianetics, flying saucers, Bridey Murphy, food and medical fads, etc. Revised, expanded In the Name of Science. "A very able and even-tempered presentation."—The New Yorker. 363pp. 5⅜ × 8. 20394-8 Pa. $5.95

ANCIENT EGYPT: ITS CULTURE AND HISTORY, J. E Manchip White. From pre-dynastics through Ptolemies: society, history, political structure, religion, daily life, literature, cultural heritage. 48 plates. 217pp. 5⅜ × 8½. 22548-8 Pa. $4.95

SIR HARRY HOTSPUR OF HUMBLETHWAITE, Anthony Trollope. Incisive, unconventional psychological study of a conflict between a wealthy baronet, his idealistic daughter, and their scapegrace cousin. The 1870 novel in its first inexpensive edition in years. 250pp. 5⅜ × 8½.   24953-0 Pa. $4.95

LASERS AND HOLOGRAPHY, Winston E. Kock. Sound introduction to burgeoning field, expanded (1981) for second edition. Wave patterns, coherence, lasers, diffraction, zone plates, properties of holograms, recent advances. 84 illustrations. 160pp. 5⅜ × 8¼. (Except in United Kingdom)   24041-X Pa. $3.50

INTRODUCTION TO ARTIFICIAL INTELLIGENCE: SECOND, EN-LARGED EDITION, Philip C. Jackson, Jr. Comprehensive survey of artificial intelligence—the study of how machines (computers) can be made to act intelligently. Includes introductory and advanced material. Extensive notes updating the main text. 132 black-and-white illustrations. 512pp. 5⅜ × 8½.   24864-X Pa. $8.95

HISTORY OF INDIAN AND INDONESIAN ART, Ananda K. Coomaraswamy. Over 400 illustrations illuminate classic study of Indian art from earliest Harappa finds to early 20th century. Provides philosophical, religious and social insights. 304pp. 6⅜ × 9⅜.   25005-9 Pa. $8.95

THE GOLEM, Gustav Meyrink. Most famous supernatural novel in modern European literature, set in Ghetto of Old Prague around 1890. Compelling story of mystical experiences, strange transformations, profound terror. 13 black-and-white illustrations. 224pp. 5⅜ × 8½. (Available in U.S. only)   25025-3 Pa. $5.95

ARMADALE, Wilkie Collins. Third great mystery novel by the author of *The Woman in White* and *The Moonstone*. Original magazine version with 40 illustrations. 597pp. 5⅜ × 8½.   23429-0 Pa. $7.95

PICTORIAL ENCYCLOPEDIA OF HISTORIC ARCHITECTURAL PLANS, DETAILS AND ELEMENTS: With 1,880 Line Drawings of Arches, Domes, Doorways, Facades, Gables, Windows, etc., John Theodore Haneman. Sourcebook of inspiration for architects, designers, others. Bibliography. Captions. 141pp. 9 × 12.   24605-1 Pa. $6.95

BENCHLEY LOST AND FOUND, Robert Benchley. Finest humor from early 30's, about pet peeves, child psychologists, post office and others. Mostly unavailable elsewhere. 73 illustrations by Peter Arno and others. 183pp. 5⅜ × 8½.
22410-4 Pa. $3.95

ERTÉ GRAPHICS, Erté. Collection of striking color graphics: *Seasons, Alphabet, Numerals, Aces* and *Precious Stones*. 50 plates, including 4 on covers. 48pp. 9⅜ × 12¼.   23580-7 Pa. $6.95

THE JOURNAL OF HENRY D. THOREAU, edited by Bradford Torrey, F. H. Allen. Complete reprinting of 14 volumes, 1837–61, over two million words; the sourcebooks for *Walden*, etc. Definitive. All original sketches, plus 75 photographs. 1,804pp. 8½ × 12¼.   20312-3, 20313-1 Cloth., Two-vol. set $80.00

CASTLES: THEIR CONSTRUCTION AND HISTORY, Sidney Toy. Traces castle development from ancient roots. Nearly 200 photographs and drawings illustrate moats, keeps, baileys, many other features. Caernarvon, Dover Castles, Hadrian's Wall, Tower of London, dozens more. 256pp. 5⅜ × 8¼.
24898-4 Pa. $5.95

AMERICAN CLIPPER SHIPS: 1833–1858, Octavius T. Howe & Frederick C. Matthews. Fully-illustrated, encyclopedic review of 352 clipper ships from the period of America's greatest maritime supremacy. Introduction. 109 halftones. 5 black-and-white line illustrations. Index. Total of 928pp. 5⅜ × 8½.
25115-2, 25116-0 Pa., Two-vol. set $17.90

TOWARDS A NEW ARCHITECTURE, Le Corbusier. Pioneering manifesto by great architect, near legendary founder of "International School." Technical and aesthetic theories, views on industry, economics, relation of form to function, "mass-production spirit," much more. Profusely illustrated. Unabridged translation of 13th French edition. Introduction by Frederick Etchells. 320pp. 6⅛ × 9¼. (Available in U.S. only)
25023-7 Pa. $8.95

THE BOOK OF KELLS, edited by Blanche Cirker. Inexpensive collection of 32 full-color, full-page plates from the greatest illuminated manuscript of the Middle Ages, painstakingly reproduced from rare facsimile edition. Publisher's Note. Captions. 32pp. 9⅜ × 12¼.
24345-1 Pa. $4.50

BEST SCIENCE FICTION STORIES OF H. G. WELLS, H. G. Wells. Full novel *The Invisible Man*, plus 17 short stories: "The Crystal Egg," "Aepyornis Island," "The Strange Orchid," etc. 303pp. 5⅜ × 8½. (Available in U.S. only)
21531-8 Pa. $4.95

AMERICAN SAILING SHIPS: Their Plans and History, Charles G. Davis. Photos, construction details of schooners, frigates, clippers, other sailcraft of 18th to early 20th centuries—plus entertaining discourse on design, rigging, nautical lore, much more. 137 black-and-white illustrations. 240pp. 6⅛ × 9¼.
24658-2 Pa. $5.95

ENTERTAINING MATHEMATICAL PUZZLES, Martin Gardner. Selection of author's favorite conundrums involving arithmetic, money, speed, etc., with lively commentary. Complete solutions. 112pp. 5⅜ × 8½.
25211-6 Pa. $2.95
THE WILL TO BELIEVE, HUMAN IMMORTALITY, William James. Two books bound together. Effect of irrational on logical, and arguments for human immortality. 402pp. 5⅜ × 8½.
20291-7 Pa. $7.50

THE HAUNTED MONASTERY and THE CHINESE MAZE MURDERS, Robert Van Gulik. 2 full novels by Van Gulik continue adventures of Judge Dee and his companions. An evil Taoist monastery, seemingly supernatural events; overgrown topiary maze that hides strange crimes. Set in 7th-century China. 27 illustrations. 328pp. 5⅜ × 8½.
23502-5 Pa. $5.00

CELEBRATED CASES OF JUDGE DEE (DEE GOONG AN), translated by Robert Van Gulik. Authentic 18th-century Chinese detective novel; Dee and associates solve three interlocked cases. Led to Van Gulik's own stories with same characters. Extensive introduction. 9 illustrations. 237pp. 5⅜ × 8½.
23337-5 Pa. $4.95

Prices subject to change without notice.
Available at your book dealer or write for free catalog to Dept. GI, Dover Publications, Inc., 31 East 2nd St., Mineola, N.Y. 11501. Dover publishes more than 175 books each year on science, elementary and advanced mathematics, biology, music, art, literary history, social sciences and other areas.